VOLUME 6

ADVANCES IN FISH
AND WILDLIFE
ECOLOGY & BIOLOGY

ABOUT THE EDITORS

Dr. Bansi Lal Kaul (b.1942) is a noted teacher, researcher, administrator and author. He has to his credit over a hundred research and popular articles on ecology and biology of fishes, Himalayan ecology and wildlife. Besides having edited titles on Himalayan biodiversity, conservation and degradation of Himalayan environment he is editing a series titled, "Advances in Fish and Wildlife Ecology and Biology". The present volume is the sixth in the series. Dr. Kaul is the recipient of several National and International honours and awards. Currently he is associated with "Solutions Exchange" programme of the United Nations in India.

Dr. Anil Verma (b.1963) Associate Professer and Head, Faculty of Sciences, Government Post-Graduate College of Education, Jammu, J&K State did his M.Sc. Zoology in 1986. He was awarded M.Phil (1988) and Ph.D (1993) degrees by the University of Jammu, Jammu for his work in the field of animal reproduction. Dr. Verma was elected a fellow of the Linnean Society of London in 2006. He has to his credit more than 70 research papers and review articles published in reputed journals. Besides teaching he guides research.

He is the recipient of Rashtriya Gourav Award of the India-International Friendship Society and the top 100 Scientists/Professionals of the year 2012 by American Biographical Institute, Washington.

ADVANCES IN FISH & WILDLIFE ECOLOGY & BIOLOGY

Volume-6

Editor in Chief
B.L. KAUL
Former Professor of Zoology and Principal,
Government S.P.M.R. College of Commerce and Management,
University of Jammu, Jammu-180001, J&K, India

Editor
A.K. VERMA
Associate Professor and Head,
Faculty of Sciences,
Government Postgraduate College of Education,
Jammu, J&K, India

2015
Daya Publishing House®
A Division of
Astral International (P) Ltd
New Delhi 110 002

Published by   :   **Daya Publishing House®**
                          A Division of
                          Astral International Pvt. Ltd.
                          – ISO 9001:2008 Certified Company –
                          House No. 96, Gali No. 6,
                          Block-C, 30ft Road, Tomar Colony, Burari
                          New Delhi-110 084
                          E-mail: info@astralint.com
                          Website: www.astralint.com

Sales Office   :   4760-61/23, Ansari Road, Darya Ganj
                          New Delhi-110 002   Ph. 011-23245578, 23244987

Laser Typesetting   :   **Shakun Graphics**
                          Delhi-110 032

Printed at   :   **Replika Press Pvt. Ltd.**

PRINTED IN INDIA

Dedicated to

LATE SAMSAR CHAND KOUL
ORNITHOLOGIST

PREFACE

The global convention on Biodiversity, signed at the earth summit in 1992 described Biological Diversity as, "the variability among all living organisms from all sources, including terrestrial, marine and aquatic ecosystems and ecological complexes of which they are part, this includes diversity within species and of ecosystems". The diversity of nature is the result of an evolutionary process that started two billion years ago. However, mankind is destroying biodiversity at an incredible speed. The vivid example before us is that of the rainforests which have been destroyed just in a period of twenty years.

The number of species endangered world over due to anthropogenic causes and natural or semi-natural habitats being destroyed, fragmented or changed are constantly growing. This is resulting in destabilization of ecosystems thereby causing the loss of vital resources together with genetic and cultural impoverishment. The pressures on Biodiversity emanate from all sectors of society, with agriculture, forestry, industry and transport being particularly responsible for habitat loss and fragmentation.

The atmosphere covering the planet Earth keeps it warm. Without the warming cover of natural greenhouse gases, mainly CO_2 and water vapour, life could not exit on the earth. Through the release of greenhouse gases such as CO_2, methane, CFC's and N_2O caused by human activities, global warming is increasing and our climate changing. There is consensus in the scientific community that climate change may result in serious consequences such as melting of glaciers, change in rainfall levels and patterns, acceleration in the loss of Biodiversity and frequent recurrence of weather anomalies such as tornadoes and hurricanes.

Being conscious of these consequences the recently finished Doha climate conference (07 /08 Dec. 2012) agreed on a second commitment period of the Kyoto protocol from 2013 to 2020. However, the second commitment period will have limited impact on emissions by 2020 as:

1. The participation of countries with emission reduction commitment is small;

2. the reduction commitments are less ambitious than needed; and

3. the allowances not used in the first commitment period can be carried over to the next commitment period where they replace actual emission reduction efforts.

As in the earlier volumes the present volume is also divided into two sections namely Fish and Limnology and Wildlife. The current volume is dedicated to a reputed bird watcher of yester years from Kashmir Mr. Samsar Chand Kaul. I am reproducing in his honour my popular write up on birds of Kashmir published on 31st October, 2010 in an esteemed and well circulated daily, "Greater Kashmir" published from Srinagar, Kashmir, India. My dear friend Prof. (Dr.) P.L. Koul, on my request, has written a tribute in the memory of Late Mr. Samsar Chand Kaul.

I hope the present volume will receive the same attention as the earlier ones. The volume would not see light of the day without active cooperation of the contributors, and encouragement and help from my family particularly my wife Promila. Last but not the least I am thankful to Mr. Anil Mittal the Publisher, for his unstinted help and cooperation as in the past in bringing out the volume in the shortest possible time.

B.L. Kaul
Jammu

MASTER SAMSAR CHAND KOUL (1883-1997) – A TRIBUTE

Master Samsar Chand Koul, a graduate with Persian-Urdu background, yet a member of National Geographical Society, Washington; Royal Geographical Society Canada and Society of World Bird Watchers England, in as early as 1950, was born in 1883 at Motiyar, Vital Sahib, Rainawari Srinagar, Kashmir (India) and married to Lachch Kuji – a Kashmiri lady.

An educationist, Environmentalist and Ornithologist of great repute, far ahead of his times, he was a teacher by profession for nearly 40 years in CMS School, Fateh Kadal, Srinagar later named Central High School, Srinagar about whom The Revd. Canon C.E. Tyndale-Biscoe a great pioneer of education in Kashmir and a missionary wrote "I am most grateful to Samsarchand for having taught his boys to love birds through his teaching of natural history".

A widely travelled person (along with his students) masterji had scaled some of the mountain peaks of Kashmir Valley namely, the Mahadev, the Harmukh and the Kolahoi. He had reached some famous glaciers like Botkul, Harbhagwan, Harmokh, Kolahai and Thajwas; passed through fine meadows of Kashmir like Allapthar, Bearam Galli, Bungus, Gurez Valley, Jamia Galli and Tosch Maidan. He had also rowed through some of the great lakes (Atrophic, Oligotrophic and Eutrophic) of Kashmir like the Gangbal, the Harnag, the Kaunsarnag, the Kolsar, the Marsar, the Sheesh Nag, the Tarsar and the Vishnusar.

For bird watching (both local and migratory) he selected the Anchar lake, Hokursar, Ferozepur nallah (Tangmarg) floating gradens of the Dal Lake and islets of dense willows and popular trees. Salim Ali, the great ornithologist of India, joined him in bird watching in the Anchar Lake and the Dal Lake and some of the birds indentified by them are as under:

Brown dipper (*Cinclus pallasii*), Blue headed rock thrush (*Monticola cinclorhynchus*), Himalayan black bulbul (*Hypsipets* spp.) Himalayan whistling thrush (*Myophoneus* spp.), Hoopoe (*Upupa epops*), Indian bush chat (*Saxicola torquata indica*), Jungle Crow (*Corvus* spp.), Kingfisher (*Alcedo* spp.), Mountain finch (*Leucostica* spp.), Sand piper (*Actitis* spp.), Stoliczka bush chat (*Sexicola macrorhynchus*), Turtle dove (*Streptopilia* spp.) and Woodpeckers.

His floral collection included *Astra falconeri, Anemone, Aquilegia* (Alpine Columbine), *Corydalis cashmeriana, Primula reptans and Primula rosea* and some of his herbal collections from high maintain ridges of Kashmir included *Artemisia, Indigofera, Salvia* and *Sassurealappa* which were preserved by him in grass made baskets and were used for curing some ailments like joint pain, skin diseases, cough and cold etc. He had built a museum in a room of his school where all his collections including bird eggs, nests and floral collections were neatly preserved and displayed. All those who visited the museum admired his efforts.

With a strong passion for outing his favourite places included Achhbal, Anantnag, Gulmarg, Khirbhawani, Nagdandi and places of special interest in ancient monuments especially Kothyarwon, Parihaspore and Martand. His work earned him a foreign travel across the Bay of Bengal to Burma in 1935. A stargazer and a student of Kashmir Shaivism masterji learnt Sanskrit and developed great acumen for scientific knowledge and natural history and authored three books namely "Beautiful valley of Kashmir and Ladakh" , "Birds of Kashmir" and "Srinagar and its Environ's" in 1940's.

It is interesting that while rest of the world was trying to understand importance of Birds, Ecology and Environment, masterji had laid the foundation of these fields in Kashmir but unfortunately the importance of his work was not recognized and nothing was done to consolidate in this field for the overall betterment of Jammu and Kashmir State, despite the fact that he had produced some of the most distinguished personalities of the state. On the contrary his erstwhile English Colleagues from C.M.S. School Srinagar appreciated his efforts and were instrumental in getting him a small wartime pension from the British Government.

The best way to honour this great Bird Watcher would be to create a "Master Samsar Chand Koul chair" in the department of Zoolgy/ Botany / Environmental sciences of our universities so that Ornithology is established as a branch of study on firm footing and Jammu & Kashmir leads the world in the field of Ornithology, Ecology, Environment and Natural History.

In writing this tribute for this great soul of our land I feel greatly honoured.

Prof. (Dr.) P.L. Kaul
Professor of Zoology and
Ex-Principal
GDC Akhnoor
37 / 2 A, Roop Nagar
Enclave, Jammu
E-mail pearaykoul@gmail.com

LIST OF CONTRIBUTORS

1. Ahmed Shahnawaz, Department of Zoology, Government Degree College for women, Sopore, Kashmir, India.

2. Almeida Victor M., Universidad Nacional Autonoma de Mexico, Research and Postgraduate Division, Av. De los Barrios No. 1, 54090 Tlalnepantla, E stado de Mexico, Mexico.

3. Araujo G.S., Department of Oceanography and Limnology, Centre of Bio-Science, Universidade Fedral do Rio, Grandodo Norte (UFRN), Natal, Brazil.

4. Bohidar K., Department of Zoology, Utkal University, Vanivihar, Bhubaneswar – 751014, Odisha, India.

5. Cavalcanti, E.T.S., Department of Oceanography and Limnology, Centre of Bio-Science, Universidade Fedral do Rio, Grandodo Norte (UFRN), Natal, Brazil.

6. Chaudhari, V., Department of Zoology, Government Degree College (Boys), Kathua, J&K, India.

7. Chellappa, T., Department of Oceanography and Limnology, centre of Bio-Science, Universidade Fedral do Rio, Grandodo Norte (UFRN), Natal, Brazil.

8. Chellappa, S., Department of Oceanography and Limnology, centre of Bio-Science, Universidade Fedral do Rio, Grandodo Norte (UFRN), Natal, Brazil.

9. Costa, E.F.S., Department of Oceanography and Limnology, centre of Bio-Science, Universidade Fedral do Rio, Grandodo Norte (UFRN), Natal, Brazil.

10. Gupta, Anuradha, Department of Zoology, G.G.M. Science College, Jammu 180002, J&K, India.

11. Gaytan-Herrera Martha L., Universidad Nacional Autonoma de Mexico, Research and Postgraduate Division, AV de los Barrios No. 1, 54090 Tlalnapantla, Estedo de Mexico, Mexico.

12. Harisha, M.N., Postgraduate Department of Wildlife Studies, Kuvempu University, Shankarghatta – 577441, Shimoga, Karnataka, India.

13. Honneshappa, K., Department of Applied Zoology, Kuvempu University, Shankarghatta – 577441, Shimoga, Karnataka, India.

14. Hosetti, B.B., Postgraduate Department of Wildlife Studies, Kuvempu University, Shankarghatta – 577441, Shimoga, Karnataka, India.

15. Ishrat, Bashir, Department of Environmental Sciences, Kashmir University, Hazratbal, Srinagar, Kashmir, India

16. Jayson, E.A., Kerala Forest Research Institute, Peechi-680653, Kerala, India.

17. Kananjia, D.R., Central Institute of Fresh water Aquaculture, Kausalyaganga, Bhubaneswar – 751002, Odisha, India.

18. Kaul, B.L., 186, Upper Laxmi Nagar Sarwal, Jammu 180005, J&K, India.

19. Kar, S., Subhadra Niwas Plot No. 7766/787 Shyampur (Near Kalinga Studio) P.O. Ghatihra, via Khandagiri, Bhubaneshwar- 751003, Dist Khurda, Odisha, India.

20. Koul, P.L., 37 / 2A, Roop Nagar Enclave, Jammu J&K, India.

21. Khajuria, Anil, Department of Zoology, Government Degree College, Ramnagar, Dist. Udhampur, J&K, India.

22. Majagi, Shashikanth, Department of Zoology, Government College, Gulburga- 585105, Karnataka, India.

23. Mishra, P., Central Institute of Fresh Water Aquaculture, Kausalyaganga, Bhubaneshwar 751002, Odisha, India.

24. Nandini, S., Universidad Nacional Autonoma de Mexico, Research and Postgraduate Division, AV de los Barrios No. 1, 54090, Tlalnapantla, Estedo de Mexico, Mexico.

25. Narain, Kanwar, Regional Research Centre, North-Eastern, ICMR, Post Box No. 105, Dibrugarh- 786001, Assam, India.

26. Oliveira R.K., Department of Oceanography and Limnology, Centre of Bio-Science, Universidade Fedral do Rio, Grandodo Norte (URFN), Natal, Brazil.

27. Pandey, A.K., National Bureau of Fish Genetic Resources, Canal Ring Road, Lucknow- 226002, India.

28. Pessoa, E.K.R., Department of Oceanography and Limnology, Centre of Bio-Science, Universidade Fedral do Rio, Grandodo Norte (UFRN), Natal, Brazil.

29. Raheela, Mushtaq, Department of Zoology Government Postgraduate College, Rajouri, J&K, India.

30. Raina, M.K., 174 / 5 Trikuta Nagar, Jammu, J&K, India.

31. Rajesh, Dogra, Director, Department of Fisheries, Govt. of Jammu & Kashmir, Jammu, India.

32. Ramirez-Garcia, P., Universidad Nacional Autonoma de Mexico, Research and Postgraduation, Av. De los Barrios No. 1, 54090, Tlalnapantla Estado de Mexico, Mexico.

33. Sarma, S.S., Universidad Nacional Autonoma de Mexico, Research and Postgraduate Division, Av. De los Barrios No. 1, 54090 Tlalnapantla Estado de Mexico, Mexico.

34. Satinder Kour, Department of Zoology, Government Postgraduate College for Women, Gandhi Nagar, Jammu, J&K, India.

35. Venkateshwaralu, M., Department of Zoology, Government Postgraduate College for Women, Gandhi Nagar, Jammu, J&K, India.

36. Verma, A.K., Department of Zoology, Government Postgraduate College of Education, Jammu, J&K, India.

37. Ximenes-Lima, J.T.A., Department of Oceanography and Limnology, Centre of Bio-Science, Universidade, Fedral do Rio, Grandodo Norte (UFRN), Natal, Brazil.

38. Yousaf, Md., Department of Environment Sciences, Kashmir University, Hazratbal, Srinagar, Kashmir, India.

Contents

SECTION-II: WILDLIFE

SECTION 1

FISH AND LIMNOLOGY

1

INFLUENCE OF AQUATIC MACROPHYTES ON THE PHYTOPLANKTON STRUCTURE AND FISH COMMUNITIES IN A SHALLOW TROPICAL RESERVOIR

N.T. Chellappa, R.K. Oliveira, E.K.R. Pessoa and S. Chellappa

ABSTRACT

Shallow freshwater tropical ecosystems have historically received less scientific attention than temperate aquatic ecosystems world over. With the quickly spreading threats imposed by climate change, the arrival of exotic species, anthropogenic eutrophication and other associated problems of the incessant demand for fresh water for consumption and irrigation, are boosting research projects around the world during the last decade. It was therefore imperative to conduct the present study in the shallow tropical Cruzeta reservoir, situated in the state of the Rio Grande do Norte, Brazil. The emergence of uncontrolled aquatic macrophyte biomass and its impact on the composition and biomass production of phytoplankton and fish assemblages were investigated. Cruzeta reservoir represents the only source of potable water supply to the region which is subjected to global warming effects. As a consequence, the surface water temperature increased and there was a significant growth of the aquatic macrophytes during 2008-2009 as compared to an earlier study period of 2004-2005. The results suggest that the changing climate and increased surface water temperature of this shallow freshwater ecosystems contributed to the dense growth of aquatic macrophytes accompanied by an improved degree of water transparency. The presence of aquatic macrophytes substantially altered the phytoplankton species composition, their biomass production and the fish communities as a consequence of competition between these communities.

INTRODUCTION

Reservoirs are among the dynamic, diverse and productive freshwater ecosystems of the world, and they provide a series of ecosystems services to human (Thornton et al., 1990). The multiple utilities of these ecosystems, particularly for the improvement of inland fisheries in semiarid northeast Brazil is an important cost-benefit project (Chellappa et al., 2006). Reservoir eutrophication is either natural or cultural, and infrequently stimulates the growth of macrophytes, which radically alters the structure and functioning of abiotic and biotic components of shalom water reservoirs (Tundisi and Tundisi, 2008). Aquatic macrophyte communities are often present along the littoral zone of freshwater water bodies harbouring, among other organisms, zooplankton, macroinvertebrates and fish fauna characterized by small species or juveniles of species. This microhabitat offers faunal biota both shelter and foraging places (Delariva, 1994; Mulderij et al., 2005). Besides, the very important role of macrophytes is their stimulating effects on water transparency by reducing sediment resuspension and nutrient levels (Scheffer, 1998). They Provide structurally complex environmental heterogeneity and stimulate numbers and types of niches (Chamber et al., 2008). Laboratory experiments shown that many macrophyte species produce allelopathic substances and inhibit phytoplankton growth (Mulderij et al., 2005) Furthermore, this microhabitat is also utilized as a spawning site by various fish species (Vazzoler and Menezes, 1992). Fish communities associated with aquatic macrophytes take advantage of the food availability in this microhabitat where the plants from a substratum in which algae and bacteria develop as epiphytes and accumulate detritus, contributing to the abundance of invertebrates (Junk, 1973).

Ecological interactions between fish communities and macrophytes in Brazil have been studied in the Amazon (Junk, 1973; Araujo-Lima et al., 1986; Soares et al., 1986) and Parana basins (Delariva, 1994; Meschiatti et al., 2000) and Parana-Sao Paulo (Agostinho et al., 1990) including phytoplankton assemblages (Fonseca and Bicudo, 2008; Sayer et al., 2010). However, studies relating to the freshwater fish communities of semi-arid ecoregion are rather limited (Chellappa and Chellappa, 2004; Rosa et al. 2005; Chellappa et al., 2009).

The present study was conducted in Cruzeta reservoir of Rio Grande do Norte State, where the last three years witnessed an exaggerated growth of floating and submerged vegetation, a situation greatly altered from the earlier study of 2004-2005 (Chellappa, et al. 2008). Therefore, it was felt imperative to study the ecological interactions between increasing biomass of macrophytes and phytoplankton composition. The objectives of the present study were to address three fold interests: (1) biomass; (2) to check the association of phytophnkton and fish communities with the co-exisisting aquactic macrophytes; (3) to determine the resource-competition for inorganic nutrient availability among them. This study also comments on the fish communities associated with the aquatic macrophytes in the Cruzeta reservoir.

METERIAL AND METHODS

The Present study was undertaken in Cruzeta reservoir, located in the Cruzeta municipality, Rio Grande do Norte State of Brazil, for 12 months from August, 2008 to July, 2009. This reservoir is shallow and the volume of water depends on the inflow of the River Sao Jose. The reservoir is situated in the Piranha-Assu hydrographic basin between the national grid line of 6°24'42" S and 36°47'23" W (Fig. 1). It is an ageing reservoir whose construction was completed in 1929 and was subjected to different degrees of trophic status upon the time scale. The reservoir covers a surface area of 748.79 ha and the maximum water holding capacity is around 35,000.000 m^3. The macrophytic vegetation observed in the Cruzeta reservoir was composed mainly of *Eichhornia crassipes, Ceratophyllum submersum, Pistia sp,* and *Nymphea sp.*

During this study, water samples were taken from the mid-point of the pelagic region of the reservoir from three depths (surface, mid-column and bottom). The mean depth of the reservoir is around 6 m. This study covered annual cycle of the dry and rainy seasons. Integrated water samples collected from the reservoir were pooled for phytoplankton analysis. Five Liter Van Dorn's bottle sampler was used throughout the study period. The field kit of WTM Multi 340i Multi-parameter probe was used in the field to measure ph, temperature, electrical conductivity and concentration of dissolved oxygen. Turbidity was analyzed with the help of standard Secchi disc. For analysis of inorganic nutrients, the water was initially filtered using Whatman GF/F filters, which was followed by analysis of nitrate (Goltermann et al., 1978), orthophosphate (APHA, 1985) and ammonium concentrations (Golternann et al., 1978).

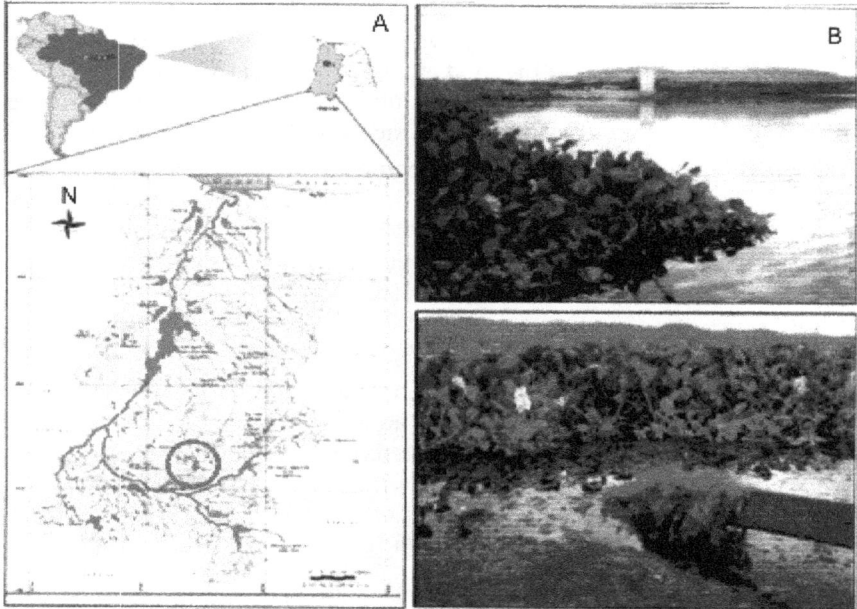

FIG. 1. (A) Study area: Cruzeta Reservoir of Rio Grande do Norte, Brazil (Area indicated within the red circle); (B) Reservoir areas with macrophytic vegetation.

Water transparency as well as total phosphorous, orthophosphate and chlorophyll a concentrations were used to calculate the Carlson trophhic status index. Information on spatial and temporal changes of the phytoplankton community was obtained from quantitative analyses of phytoplankton and chlorophyll a. Water samples were packed and protected from the light. Filtration was done through 0.7 mm fitters packed and protected from the light. Filtration was done through 0.7 mm Whatman filters (GF/F) to estimate total chlorophyll a. After filtration, the pigments were extracted with 90% acetone at 4°C in the dark. The samples were analyzed on a Libra S6 spectrophotometer (Biochrom), at wavelengths of 665 nm and 750 nm. The absorption values corresponding to the analysis of chlorophyll a were later inserted in the formula described by Marker et al. (1980), to obtain its final concentration in $ugL^{-1.}$

Phytoplankton samples were preserved in 4% Lugol-formaldeyde solution in the field. They were sedimented and analyzed using an optical microscope down to the species level or to the highest possible taxonomic resolution using the specific literature (Wehr and Sheath, 2003; Bicudo and Menezes, 2005). Phytoplankton individuals were counted through inverted microscope.

Fish samples from the reservoir were captured on a diurnal and nocturnal basis during the study period, with the help of artisanal fishermen of the region, using fishing gear which consisted of stationary nets and gillnets of different mesh sizes, cast nets, hooks and traditional traps. Random fish subsamples were collected from a larger catch landed by fishermen and were numbered and transported to the laboratory on ice. Morphometric measurements and meristic counts were carried out to check the taxonomical status of each species.

One-way ANOVA was used to test the significance of the environmental variables and the relative abundance phytoplankton assemblages.

RESULTS AND DISCUSSION

Annual rainfall in the study area was 350-650 mm during 2008-2009. The rainy season was from April to May and the dry season extended from November to February. In comparison, the earlier study during 2004-2005 registered low levels of rainfall (Fig. 2). The pH of the reservoir water oscillated from neutral to alkaline values, during both study periods. Higher water temperatures were registered during 2008-2009 as compared to the early period of study. Increase in water temperature was related to the decrease in water density. This is significant in relation to the shallow waters of these ecosystems (Fig. 3). Electrical conductivity showed higher

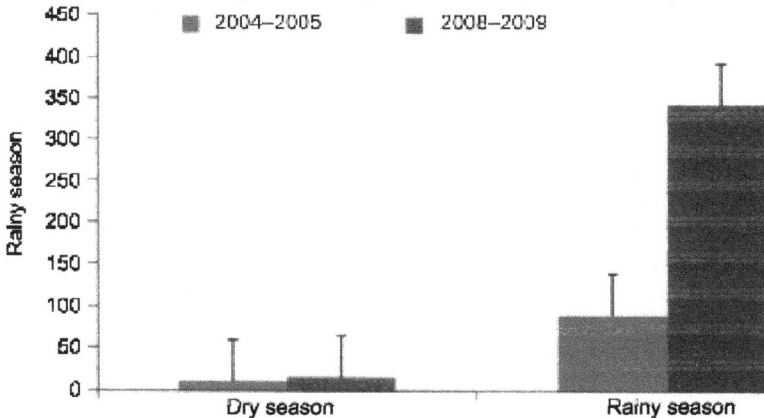

FIG. 2.   Comparison of annual rainfall pattern in the study region during the periods 2004-2005 and 2008-2009.

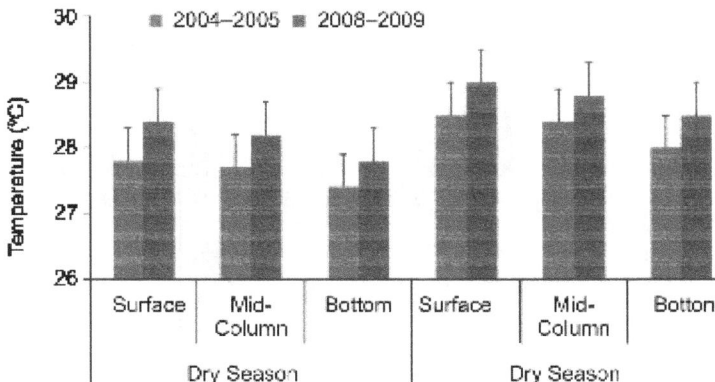

FIG. 3.   Comparison of water temperature in the study region during the periods 2004-2005 and 2008-2009.

values for dry season as compared to 2004-2005, but this trend was reversed during the rainy season. Higher biomass of macrophytes observed during 2008-2009, resulted in improved transparency of the water column, in comparison to 2004-2005 data. During the earlier study, the presence of marcophytes was negligent (Fig. 4).

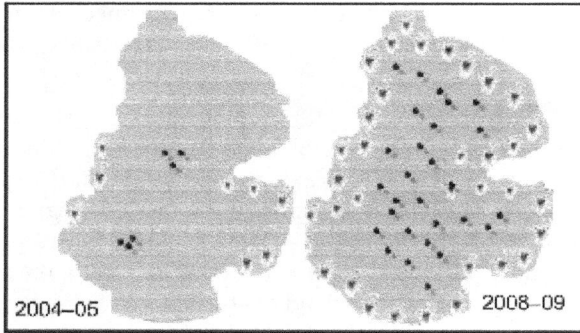

FIG. 4.   Schematic comparison of occurrence of macrophytes during the two study periods.

The distribution of dissolved oxygen concentration showed heterogeneous distribution in a vertical profile, more at surface water than at mid-column and bottom, during both study periods. The dissolved inorganic nutrient levels were higher in relation to nitrate and phosphate as compared to ammoniacal nitrogen (Fig. 5). There was a significant difference (ANOVA $p < 0.05$) in the nutrient values between the two study periods on a temporal scale. Aquatic macrophytes tend to remove selectively more ammoniacal form of nitrogen, than nitrate-nitrogen or solube reactive phosphorus. This characteristic was observed during the present study, where the high nutrient availability did not promote phytoplankton diversity instead favored the luxuriant growth of Cyanobacterial species.

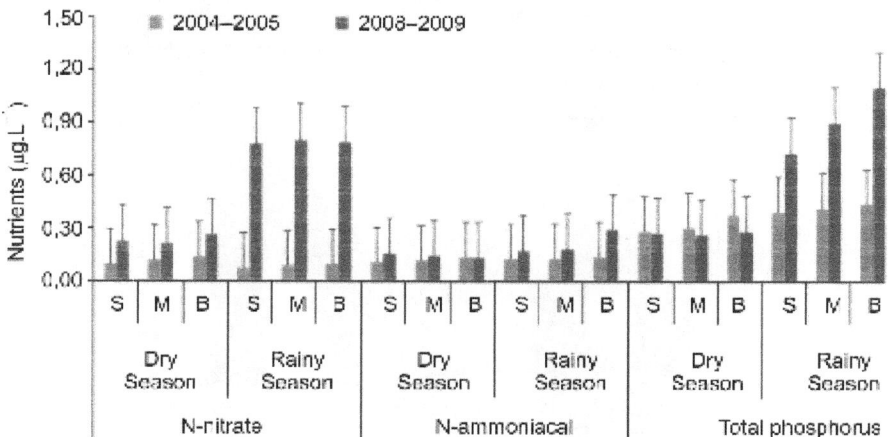

FIG. 5.   Variation of nutrients between the two study periods of 2004-2005 and 2008-2009 (S= Surface: M = Mid column; B= Bottom).

Chlorophyll *a* concentration, a functional component of ecosystems, registered a significant decline during 2008-2009 as compared to the 2004-2005. This was due to the interespecific resource competition with increased biomass production of macrophytes. This decline was marked during both seasons along a vertical profile (Fig. 6).

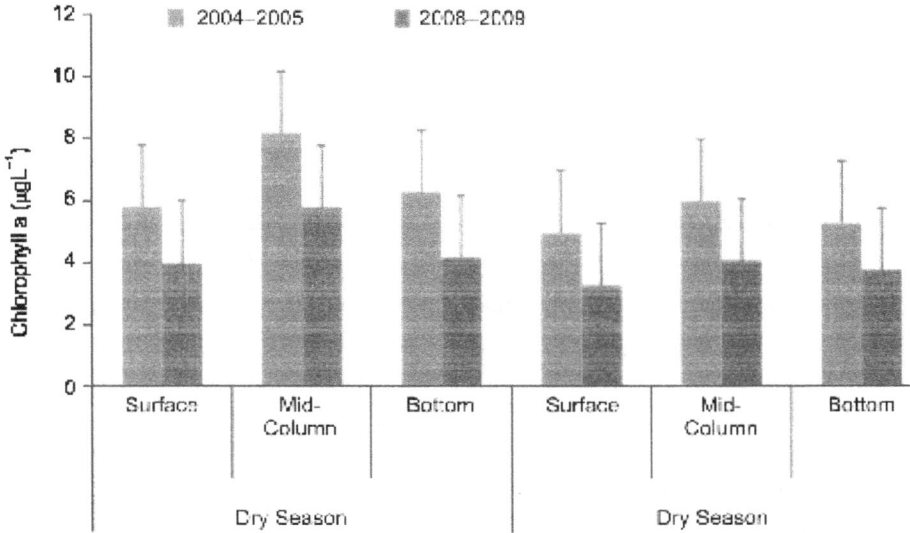

FIG. 6.   Variation of chlorophyll a between the two study periods of 2004-2005 and 2008-2009.

The results indicate the phytoplankton species composition during 2004-2005 and 2008-2009. Forty five (45) taxa were identified, which were distributed into 6 classes (Cyanobacteria, Chlorophyceae, Bacillariophyceae, Dinophyceae, Chrysophyceae and Euglenophyceae) in dry season and 5 classes with 53 taxa during the rainy season of 2004-2005. The class with the highest number of species was chlorophyceae (21 species), followed by Bacillariophyceae and Cyanobacteria. A total of 37 and 53 taxa were identified for the dry and rainy season during 2008-2009, which showed a reduction of 16.6% and an increase 3.8% over a temporal scale. There was an overwhelming dominance of cyanobacterial species representing 93.9% in dry season and 68.5% in rainy season for the period of 2008-2009 period. This depicts that the cyanobacterial species assume the role of superior competitor. Two important species of filamentous cyanobacterial species, such as, *Cylindrospermopsis raciborski* and *Leptolyngbya geophilla* were more frequent and abundant in the 2008-2009 (Fig. 7).

The present study shows the increased biomass of macrophytes and the relative abundance of cyanobacterial species, which is a response to the warming effect of temperature and the successful outcome of resource competition over phytoplankton species for the years 2008-2009 (Fig. 4). Furthermore, the species Cyanobacterium, *Cylindrospermopsis raciborskii* was considered an invasive species well adapted to wide range ecosystems and geographical distribution (Padisak, 1997). In the present study this species showed quantitative superiority in both seasons.

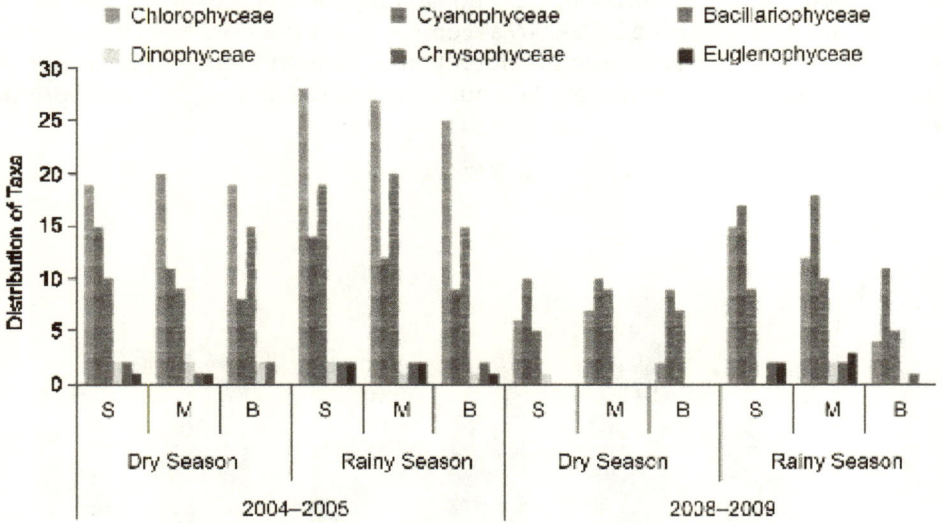

Fig. 7. Algal taxa identified during the dry rainy season of the two study periods of 2004-2005 and 2008-2009 (S= Surface; M= Mid column; B= Bottom).

There was an agreeable consensus among ecologists that biodiversity is a key factor in ecology and directly related to the regulation and functioning of the ecosystems. Both abiotic and biotic factors contributed to the variability of phytoplankton diversity at different scales with space and time. However, the present study reveals that neither transparency of water or chlorophyll a value nor phytoplankton species diversity correlated to light limitation condition (Fig. 8).

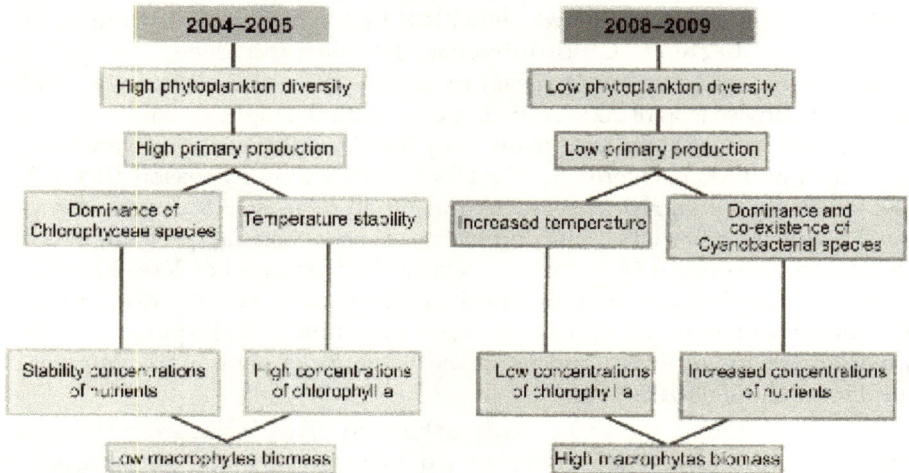

FIG. 8. Comparison of events during the two different study periods in the Cruzeta reservoir, RN, Brazil.

Carlson Trophic Status Index detected meso-eutrophic condition owing to the the increased levels of inorganic nutrients to facilitate process related to the water management.

The fish were distributed in 4 orders (Characiformes, Perciformes, Siluriformes and Synbranchiformes), 7 families (Anostomidate, Prochilodontidae, Erythrinidate, Cichlidae, Loricariidae, Synbranchidae) presenting 9 fish species (Fig. 9), of which 8 are endemic species of this ecoregion and one (*Oreochromis* niloticus) is an exotic Nile tilapia, *O. niloticus* outstands in production among the other fish species in the study reservoir. Since the beginning of 1970, *O. niloticus* was introduced in the freshwater ecosystems of the Northeastern Brazil in order to upgrade the fish production and fish culture and presently, it is a dominant species in most of the reservoirs. This is due to its rapid growth, omnivorous food habits, high reproductive efficiency, parental care and resistance to environmental variations (Zambrano et al. 2006). The macrophytes give refuge to the fish communities and also act as feeding and spawning areas for them Casatti et al., 2003).

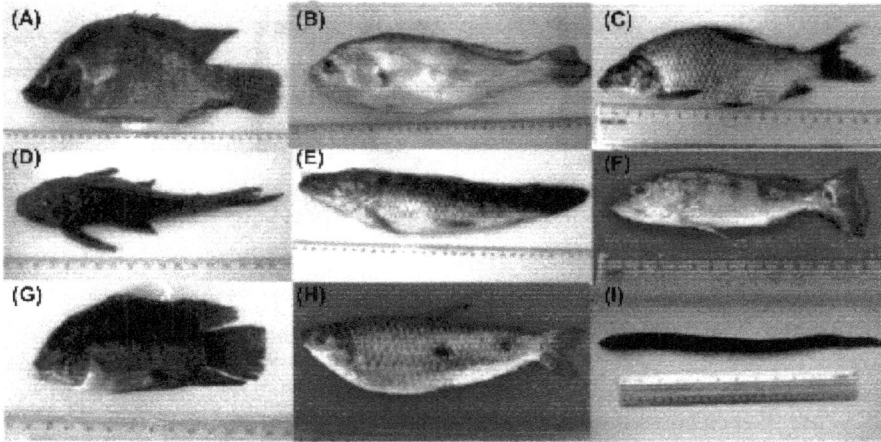

FIG. 9.   Fish communities associated with the macrophytes in the Cruzeta reservoir (A) *Oreochromis niloticus* (Linnaeus, 1758); (B) *Plagioscion squamosissimus* (Heckel, 1840); (C) *Prochilodus brevis* (Steindachner, 1875); (D) *Hypostomus pusarum* (Starks, 1913); (E) *Hoplias malabaricus* (Bloch, 1794); (F) *Cichla monoculus* (Spix & Agassiz, 1831); (G) *Cichlasoma orientale* (Kullander, 1983); (H) *Leporinus piau* (Fowler, 1941); (I) *Synbranchus marmoratus* (Bloch, 1975).

CONCLUSIONS

Macrophytes have the capacity to buffer the freshwater ecosystems against harmful consequences of anthropogenic eutrophication by suppressing phytoplankton biomass through direct and indirect mechanisms. Our study revealed that presence of macrophytes increased the transparency of water, Cyanobacterial phytoplankton and fish species composition. The diversity of phytoplankton and the chlorophyll biomass was considerably reduced. Although cyanobacterial species exist, including *Cylindrospermopsis raciborskii* their presence signifies a harmonious co-existence with macrophytes sharing nutrient resoures. The trophic index state was maintained as meso-eutrophic throughout the study period. The maintenance of macrophytes is an important target in water management of shallow water reservoirs of Rio Grande do Norte State of Northeastern Brazil to facilitate the reversal of anthropogenic eutrophication impacts.

ACKNOWLEDGMENTS

The authors wish to thank the National Council for Scientific and Technological Development of Brazil (CNPq) for the financial support awarded during the study period (N. T. Chellappa and S. Chellappa) and the Federal Post-Graduate Agency of Brazil (CAPES/MEC) (Scholorships granted to R. K. Oliveira and E. K. Pessoa). The authors also wish to thank Mr. Wallace Silva do Nascimento, of the Ichthyology laboratory of UFRN, Brazil, for organizing the figures.

REFERENCES

Agostinho, A.A., Julio Jr., H.F. and Borghetti, J.R., (1992). Consideracoes sobre os impactos dos represamentos na ictiofauna e medidas para sua atenuacao. Um estudo de caso: Reservatorio de Itaipu. Revista UNIMAR, Maringa, 14 (Suplemento): 89-107.

APHA, American Public Health Association, (1985). Standard methods for examination of water and wastewater. 16 ed. p. 1527.

Araujo-Lima, C.A., Portugal, L.P.S. and Ferreira, E.G., (1986). Fish-macrophytes relationship in the Anavilhanas Archipelago, a black water system in the central Amazon. Journal of Fish Biology, 29: 1-11.

Bicudo, C.E.M., Menezes, M., (2005). Generos de algas de aguas continentais do Brasil: chave para identificacao e descricoes. Sao Carlos: Rima, p. 489.

Casatti, L.H.F., Mendes, K. Ferreira, M., (2003). Aquatic macrophytes as feeding site for small fishes in the Rosana Reservoir, Paranapanema River, Southeastern Brazil. Brazilian Journal of Biology, 63 (2): 213-222.

Chambers, P.A., Lacoul, P., Murphy, K.J. and Thomaz, S.M., (2008). Global diversity and aquatic macrophytes. Hydrobiologia, 595: 9-26.

Chellappa, S. and Chellappa, N.T., (2004). Ecology and reproductive plasticity of the Amazonian cichlid fishes introduced to the freshwater ecosystems of the semi-arid Northeastern Brazil. In: Advances in Fish and Wildlife Ecology and Biology. Ed. B. L. Kaul. Daya Publishing House , Delhi, India. 3: 49 – 57.

Chellappa, N. T., Borba, J.L.M. and Rocha, O., (2008). Phytoplankton community structure and physical-chemical characteristics of water in the public reservoir of Cruzeta, RN, Brazil. Brazilian Journal of Biology, 68 : 477-494.

Chellappa, N.T., Chellappa, T., Lima, A.K.A., Medeiros, J.L., Souza, P.V.V. and chellappa, S., (2006). Ecology of freshwater phytoplankton assemblages from a tropical of Northeastern Brazil. International Journal of Lakes and Rivers, 1 (1): 61-81.

Chellappa, S., Bueno, R.M.X., Chellappa, T., Chellappa, N.T., VAL, V.M.F.A., (2009). Reproductive seasonality of the fish fauna and limnoecology of semi-arid Brazilian reservoirs. Limnologica 39: 325-329.

Delariva, R.L., (1994). Icthyofauna associated to aquatic macrophytes in the upper Parana river floodplain. Ecological Perspectives. Revista UNIMAR, Maringa. 16 (3):41-60.

Fonseca, B.M. and Bicudo, C.E.M., (2008). Phytoplankton seasonal variation in a shallow stratified eutrophic reservoir (Garcas Pond, Brazil). Hydrobiologia, 600: 267-282.

Golterman, H.L., Clymo, R.S. and Ohnstat, M.A.M. (1978). Methods for physical and chemical analysis of Freshwaters. IBP Handbook, Blackwell Sci. Publ. Oxford. p. 215.

Junk, W.J., (1973). Investigations on the ecology and production-biology of the "floating-meadows" (Paspalo-Echinochloetum on the Middle Amazon). II – The aquatic fauna in the root-zone of floating vegetation. Amazoniana, IV: 9-102.

Marker, A.F.H., Nusch, E.A., Rai, H. and Riemann, B., (1980). The measurements of photosynthetic pigments in freshwater and standardization of methods: conclusions and recommendations. Archives of Hydrobiologia, Beih., 14: 91-106.

Meschiatti, A.J., Arcifa, M.S. and Fenerich-Verani, N., (2000). Fish communities associated with macrophytes in Brazilian food plain lakes. Environmental Biology of fish, Dordrecht. 58 (2): 133-143.

Mulderij, G., Mooij, W.M. and Van Donk, E., (2005). Allelopathic growth inhibition and colony formation of the Green alga *Scenedesmus obliquus* by the aquatic macrophyte Stratiotes aloides. Aquatic Ecology. 39: 11-12.

Padisak, J. (1997). *Cylindrospermopsis raciborskii* (Woloszynnska) Seenayya et Subba Raju, an expanding, highly adaptive cyanobacterium: worldwide distribution and review of its ecology, Archiv fur Hydrobiology. 107: 563-593.

Rosa, R.S., Menezes, N.A., Britski, H.A., Costa, W.J.E.M. and Groth, F., (2005). Diversidade, padroes de distribuicao e conservacao dos peixes da Caatinga In: Leal, I.R., Tabarelli, M. and Silva, J.M.C. eds. Ecologia e Conservacao da Caatinga. Editora UFPE, Recife, 135-180.

Sayer C.D., Burgess, A., Kari, K., Davidson, T.A., Peglar, S., Yange, H. and Rose, N., (2010). Long-term dynamics of submerged macrophytes and algae in a small and shallow, eutrophic lake: worldwide distribution and review of its ecology. Archiv fur Hydrobiology. 107: 563-593.

Rosa, R.S., Menezes, N.A., Britski, H.A., Costa, W.J.E.M. and Groth, F., (2005). Diversidade, padroes de distribuicao dos peixes da Caatinga. In: Leal, I.R., Tabarelli, M. and Silva, J.M.C. eds. Ecologia e Conservacao da Caatinga. Editora UFPE, Recife, 135-180.

Sayer C.D., Burgess, A., Kari, K., Davidson, T.A., Peglar, S., Yang, H. and Rose, N., (2010). Long-term dynamics of submerged macrophytes and algae in a small and shallow, eutrophic lake: implications for the stability of macrophyte-dominance. Freshwater Biology. 55: 565-583.

Scheffer, M., (1998). Ecology of Shallow Lakes. Chapman & Hall, London, p. 357.

Soares, M.G.M., Almeida, R.G. and Junk, W.J. (1986). The tropic status of the fish fauna in lago Camaleao e macrophyte dominated Floodplain Lake in the middle Amazon. Amazoniana, 9 (4): 511-526.

Thornton, K.W., Kimmel, B.L. and Payne, F.E. (eds), (1990). Reservoir Limnology: ecological perspectives. New York: John Wiley and Sons, 15-41.

Tundisi, J. G. and Tundisi, T. M. 2008. Limnologia. Sao Paulo: Oficina de Textos.

Vazzoler, A.E.A.M. and MENEZES, N.A. (1992). Sintese de conhecimentos sobre o comportamento reprodutivo dos Characiformes da America do Sul (Teleostei, Ostariophysi). Revista Brasileira de Biologia. 52(4): 627-640.

Wehr, J.D. and Sheath, R.G. (2003). Freshwater algae of North-America: ecology and classification. London: Academic Press. p. 918.

Zambrano, L., Martinez-Meyer, E., Menezes, N.A. and Petersen, A.T., (2006). Invasive potential of common carp (*Cyprinus carpio*)and *Nile tilapia* (Oreochromisniloticus) in American freshwater systems. Canadian Journal of Fisheries and Aquatic Science 63: 1906- 1910.

□□□

2

COMPARATIVE ASPECT OF SEED PRODUCTION OF *MACROBRACHIUM GANGETICUM* (BATE) IN NATURAL SEAWATER AND BRINE SOLUTION

Prasanti Mishra, D. R. Kanaujia, K. Bohidar and A. K. Pandey

ABSTRACT

Larval rearing of *Macrobrachium gangeticum* was carried out under hatchery conditions using natural seawater and concentrated brine solution for two successive years. Both the media were prepared by diluting with freshwater to maintain 12 ppt salinity. First zoea larvae were stocked @ 100 larvae/L in 300 L plastic tanks adopting airlift biofilter re-circulatory system. Larvae were fed with *Artemia nauplii* twice daily for one week, thereafter egg custard and mussel meat at six hourly intervals including *Artemia nauplii* during night. Desired water quality parameters were maintained throughout the rearing period. Larvae passed through eleven distinct larval stages and first post-larvae (PL) appeared between day 19 and 22 in both media and cycle was completed on day 40. The production of PL ranged from 8,726-9,009 with density of 29-30 PL/*l* during first year and 7,744-8,905 with density of 26-30 PL/*l* during second year in natural sea water. Almost similar trend in PL production was recorded in brine solution too as it ranged form 6,498-7,825 with density 22-26 PL/*l* during first year and 6,00-7,096 with density of 20-24 PL/*l* during second year. Though slightly higher production of PL was observed in the natural seawater, brine solution may be used in the operation of inland hatchery of PL production of Gangetic prawn.

Keywords: *Larval rearing, Brine solution, Seed production, Macrobrachium gangeticum.*

INTRODUCTION

The giant freshwater prawn (*Macrobrachium rosenbergii*) farming has taken the shape of an industry globally ever since the life-cycle was established (and closed) in Malaysia by Ling (1969). A few countries like Thailand and Taiwan took advantage of this breakthrough and started prawn farming early while India and Bangladesh initiated the prawn farming very recently (MPEDA, 2001; FAO, 2002a). The two International Symposia on Freshwater Prawns organized in Cochin, Kerala (India) during August 2003 and January 2011 strongly emphasized the vital need to augment quality seed production to improve sustain freshwater prawn culture throughout the world. For instance the freshwater prawn production potential of India is 1,50,000 tones worth Rs. 3,000 cores @ Rs. 200/kg (US$ 65 million). Among the nine coastal states of India, Andhra Pradesh takes the lions share (88.6%) of production of freshwater prawns with productions of about 27,020 tones (MPEDA, 2004).

Drastic changes in water qualities have been recorded with progression of larval rearing and its maintenance is the key for successful seed production of prawns under hatchery conditions. Brackishwater of desired salinity used for hatchery operation, is prepared by diluting seawater with freshwater though it contains varying concentrations of impurities including colloidal suspensions and dissolved organic matter (New and Singholka, 1985; Kanaujia et al., 2005; Kanaujia, 2006; New and Valenti, 2009). Therefore, chemical treatments, followed by aging, are essential for purification of water before use in prawn hatchery (Kanaujia *et al.*, 2007). *Macrobrachium gangeticum* is the third largest freshwater prawn recorded from river Ganges and Brahmaputra flowing through Uttar Pradesh, Bihar, Bengal, Assam and Arunachal Pradesh. This Species is reported to migrate up to 1,300 km from estuary to freshwater (riverine) system where they grow, mature, breed and spawn. Larval development naturally occurs in brackishwater in the estuary (Kanaujia *et at.*, 2001, 2005, 2007; Kanaujia *et al.*, 2007). Hence development of hatchery needs large quantity of brackishwater for commercial seed production which is labour-intensive and expensive and is viable only in coastal regions with use of natural seawater. Brine obtained from salt pans diluted with well or surface water is being used for larval rearing of *M. rosenbergii* in Thailand (Chowdhury *et al.*, 1993; Hangsapreurke *et al.*, 2008). An attempt has been made to evaluate and compare the possible use of brine solution and natural seawater for seed production of *M. gangeticum* to establish hatcheries in inland areas located far away from the coastal regions.

MATERIAL AND METHODS

The experiments were conducted at Central Institute of Freshwater Aquaculture (CIFA), Kausalyaganga, Bhubaneshwar (Orissa) (latitude 2011'6"-20 11'45" N; longitude 85 50' 52"- 85 51'35" E) using 35 ppt natural seawater transported from Puri seashore to Prawn Hatchery Complex, CIFA and stored in a plastic pool (4'hx16' diameter) kept under sunlight exposure. Similarly, 150 ppt concentrated brine solution was transported from Huma Salt Farm, Ganjam District (Orissa). Both the water media were diluted with freshwater to obtain 12 ppt desired level of salinity for larval rearing and treated with 5 mg/1 sodium hypochloride (NaOC*l*) and agitated vigorously for two days for through electic agitator, left to settle down the suspended particles and was kept for 15 days for growth of phytoplankton to reduce the concentration of dissolved nutrient load. Once the phytoplankton blooms appeared and settled at the bottom and sides of the tank, the clear water was pumped out and filtered again through a sand filter and treated again with 5mg/1 sodium hypo chloride and repeated the above process 3 times to get the clear water medium for larval rearing. If the residual chlorine was observed in the medium, it was neutralized by adding sodium thiosulphate ($Na_2S_2O_3H_2O$) at 1:5 by weight on the following day and taken to the hatchery after filtering again through a sand filter (0.5 m bed). The brood prawns were maintained in hatchery under captivity using 5 ppt brackishwater along with proper aeration and food. The berried females carrying grey eggs were collected and reared in half filled larval rearing tank with 5 ppt brackishwater. Soon after hatching, salinity of the medium was increased gradually by adding hyper-saline water to increase salinity from 5 to 12 ppt as required for the rearing of Macrobrachium gangeticum larvae. The larvae present in the tank were assessed through 10 randomized samples taken in 100 ml beaker and counted one-by-one by pouring the water. Same technique was adopted to assess the larval density during the entire larval-rearing cycle.

Larvae were fed with freshly hatched *Artemia* nauplii twice daily between 6-7 h and 17-18 h for one week. Thereafter, the feed was supplemented with mussel meat and egg custard at an interval of six hours using *Artemia* nauplii between 23-24 h. Once few post-larvae appeared in the tank within 19-22 days, small amount of mussel meat and/or egg custard was provided at every two hours interval till the rearing was completed. Tanks were cleaned daily during morning by siphoning off water along with the metabolites, moulted shells and left-over feed. Thereafter, water column and quality were maintainted and feed was provided. Water quality parameters like temperature, salinity, pH, ammonia, total alkalinity and dissolved oxygen were analyzed at regular intervals following standard methods (APHA, 1985).

As soon as a few first post-larvae appeared, strings shells designed at CIFA (Kanaujia *et al.*, 2002) were hung into the tanks. The strings shells were lifted carefully and kept in a tub containing 61 water of the same tank and then the post-larvae hidden in between the shells came out form the shell bed and started moving. In this process, newly metamorphosed post-larvae were removed daily. Since the post-larvae developed in higher salinity, they were acclimatized gradually in freshwater, counted and released in a freshwater tank for juvenile production.

RESULTS AND DISCUSSION

M. gangeticum larvae were transparent/translucent with red and blue chromatophores during early stages. The colour deepened at later stages of development and was confined only on some portions of the body. Larvae of all the eleven stages were active swimmer's planktonic in nature, photopositive (attracted towards light) and displayed churning movement during early stages and moved along the side of the tank and water column at later stages. They were active and moved up side down (tail up and head down) obliquely into the water column in the tank. Initially, larvae readily accepted *Artemia* nauplii but later showed feeding propensity towards egg custard and mussel meat.

Water Quality Parameters

Temperature: Variations in water temperature in different experimental units recorded during two years are presented in Table 1 and Fig. 1a,2a. It varied from 19.0-30.0°C with an average of 29.4±0.33°C (Fig. 1a) recorded during first year in natural seawater (medium1) and slight More or less similar water temperature values recorded in brine solution during two years which ranged from 29.0–30.0°C with an average of 29.4±0.33°C (Fig. 2a) during first year and 29.5–30.2°C with an average of 29.9±0.22°C during second year.

Salinity: The salinity ranged from 12–16 ppt with an average of 14.5±1.50 ppt in first year and 10-15 ppt with an average of 13.2±1.54 ppt during second year in natural sea water (medium 1) (Table 1; Fig. 1b). It ranged from 12–16 ppt with an average of 14.3±1.37 ppt during first year and 10-16 ppt with an average of 13.3±1.67 ppt during second year in brine solution (medium 2) significant variations in both the media during two years.

Table 1: Mean Physico-chemical Parameters of 3 Larval Rearing Trials in Medium 1 (Natural Seawater) and Medium 2 (Brine Solution) During Two Years

Medium 1 Natural Seawater

Parameters	2003				2004				
	Trial 1	Trial 2	Trial 3	Average	Trial 1	Trial 2	Trial 3	Average	Average total
Temperature	29.4±350	29.4±.350	29.9±0.23	29.4	29.4±350	29.9±231	29.9±0.23	29.9	29.65
Salinity	14.5±1.643	14.6±1.50	13.5±1.22	14.4	14.3±1.63	13±1.78	13±1.78	13.1	13.81
pH	7.7±0.116	7.7±0.081	7.68±0.09	7.7	7.78±0.104	7.6±0.10	7.6±0.08	7.62	7.67
Dissolved oxygen	4.3±0.18	4.3±0.179	4.27±0.16	4.3	4.2±0.178	4.4±0.23	4.3±0.08	4.3	4.29
Total hardness	2246±10.2	2255±18.7	2282±5.79	2254	2262±17.74	2264±13.3	2285±8.09	2410.16	2265.5
Total alkalinity	85.5±4.97	85.5±4.273	82.4±5.28	85.5	85.6±5.259	90±1.60	87.9±0.91	87.7	86.15
Ammoniacal nitrogen	0.088±0.007	0.078±005	0.115±005	0.082	0.081±.004	0.128±.006	0.107±005	0.116	0.0995

Medium 2 Brine solution

Parameters	2003				2004				
	Trial 1	Trial 2	Trial 3	Average	Trial 1	Trial 2	Trial 3	Average	Average total
Temperature	29.4±0.350	29.4±.350	29.9±0.21	29.4	29.4±0.350	29.9±0.231	29.9±.231	29.9	29.65
Salinity	14.6±1.50	14.3±1.36	13.3±1.11	14.3	14±1.41	13.3±2.160	13.4±1.78	13.3	13.81
pH	7.7±0.11	7.7±.081	7.7±0.09	4.23	4.2±0.17	4.4±0.233	4.3±0.081	4.33	4.28
Dissolved oxygen	4.2±0.183	4.3±.178	4.3±0.14	4.23	4.2±0.17	4.4±0.233	4.3±0.081	4.33	4.28
Total hardness	2246±10.2	2255±18.7	2276±5.2	2254	2262±14.7	2264±13.3	2284±8.09	2274	2264.5
Total alkalinity	85.8±4.97	86±4.27	86.9±4.82	86	86.5±5.28	90±1.60	87.9±0.919	88.2	87.18
Ammoniacal nitrogen	.088±.007	.078±.005	0.1±0.004	.082	0.081±0.004	.128±.006	0.107±.005	0.111	0.097

pH: The pH of larval rearing medium in different trials was maintained through the adoption of the airlift bio-filter re-circulatory system along with management of water exchange and applicsation of calcium sulphate and Na-EDTA. The variations in pH was increased initially but remained lower after fourth week onwards which was recorded more or less similar in both the media. The variations in pH ranged from 7.5-7.9 with an average of 7.7±0.10 and more of less similar pH value 7.5-7.8 with an average of 7.7± 0.09 recorded during first and second year respectively in natural seawater (medium 1) (Table 1; Fig. 1c). It ranged from 7.5-7.9 with an average of 7.7±0.10 during first year and 7.5-7.8 with an average value 7.6±0.091 during second year in brine solution (medium 2) (Table 1; Fig. 2c).

Dissolved oxygen: Due to the brackishwater environment, value of dissolved oxygen recorded in 2 media in different weeks was over 4 mg/1 during entire cycle in spite of continuous aeration which was found similar in both the media. DO ranged from 4.0-4.5 mg/l with an average of 4.2±0.17 mg/l during first year and recorded slightly higher range 4.2-4.8 mg/l with an average of 4.2±0.17 mg/l during first year 4.0-4.8 mg/l with an average of 4.3±0.17 mg/l during second year in medium 2 (Table 1; Fig. 2d).

Total hardness: Mean value of total hardness in the two test media during different weeks recorded gradual increase with the progress of larval cycle. It varied from 2,230-2,290 mg/l with an average of 2,276.5±15.71 mg/l during first year and slightly higher range 2,240-2,290 mg/l with an average of 2,276.5±12.80 mg/l was recorded during second year in medium I (Table 1; Fig 1e). The value ranged from 2,230-2,280 mg/l with an average of 2,254.4±15.71 mg/l during first year and it varied from 2,240-2,290 mg/l with an average 2,276.5±12.83 mg/l during second year in medium 2 (Table 1; Fig. 1e).

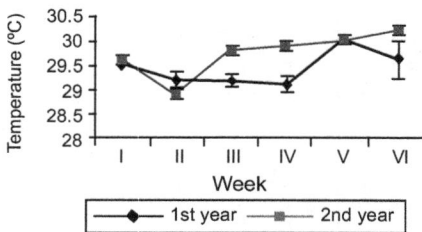

(a)

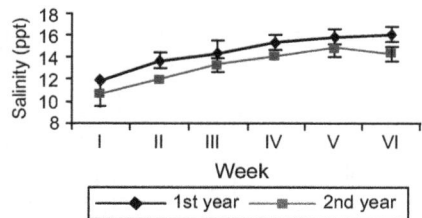

(b)

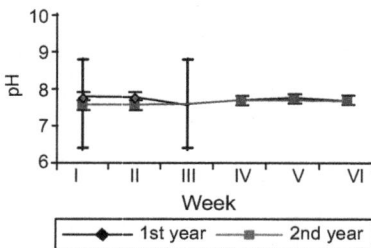

(c)

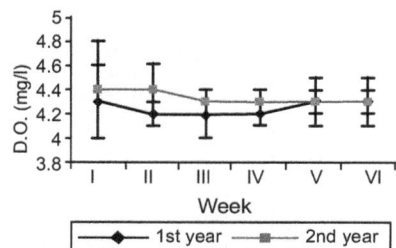

(d)

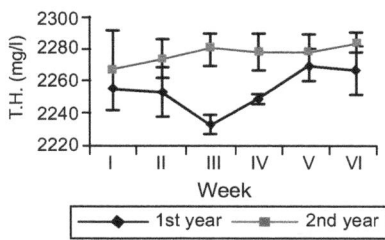

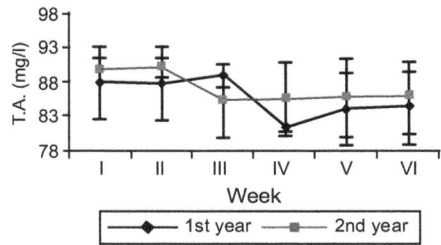

(e)   (f)

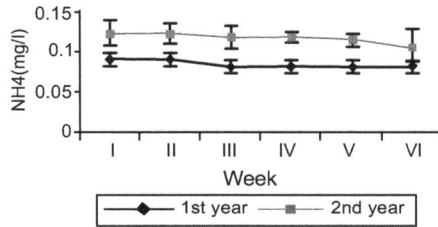

(g)

FIG. 1.(a to g)   Weekly variations in important water quality parameters in larval rearing trials of medium 1 (natural sea water) in two years.

Total alkalinity: Variations in total alkalinity ranged from 80.7-91.4 mg/l with an average 85.7±4.56 mg/l during first year and 79.2-91.7 mg/l with an average 86.9±4.32 mg/l during second year in medium 1 (Table 1 ; Fig. f). It ranged from 807.91.4 mg/l with an average 85.7±4.56 mg/l during first year and 79.2-91.7 mg/l with an average 86.9±4.32 mg/l during second year in medium 2 (Table 1; Fig. 2f). Variation in total alkalinity level of the different media was recorded to be trace and negligible.

Ammoniacal nitrogen: The dissolved ammonia ranged from 0.075-0.098 mg/l with an average 0.082±0.007 mg/l in three trials during first year and 0.099-0.137 mg/l with an average 0.1±0.01 mg/l during second year in medium 1 (Table 1; Fig. 1g). It was recorded 0.072-0.098 mg/l with an average 0.082±0.007 mg/l during first year and 0.099-0.137 mg/l with an average 0.11±0.01 mg/l during second year in medium 2 (Table 1; Fig. 2g). Maximum value recorded during second year was significantly higher than first year. There was no accumulation of ammonia in the test media during both the years.

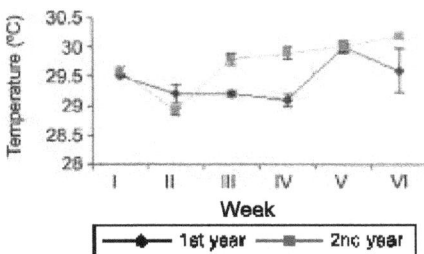

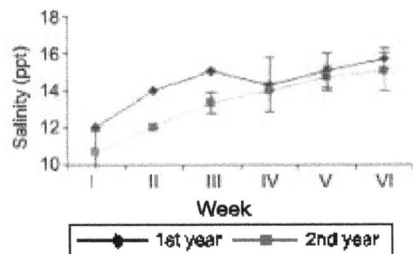

(a)   (b)

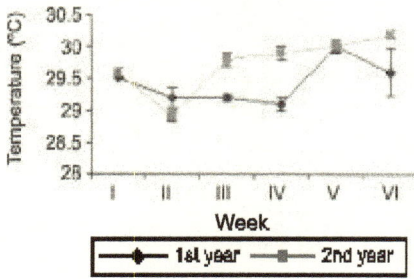

(c)

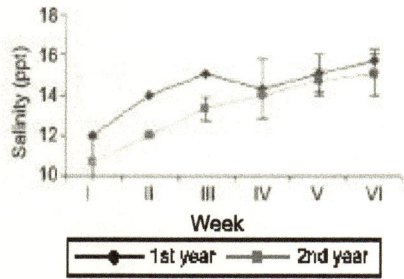

(d)

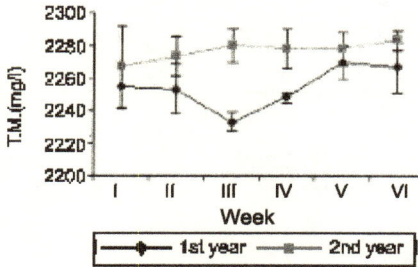

(e)

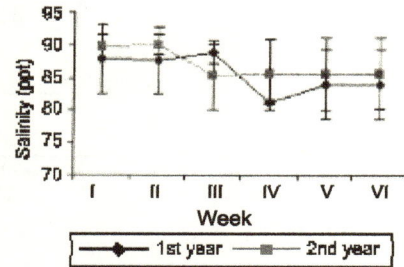

(f)

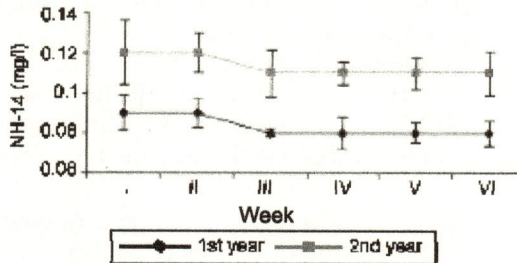

(g)

FIG. 2.(a to g) Weekly mean variations in important water quality parameters in larval rearing trials of medium 2 (brine solution) during two years (2003–2004).

Post-larvae (PL) production: The production of PL in medium 1 (natural sea water) ranged from 8,726-9,009 PL with an average 8,893.6 PL during first year and comparatively lower 7,744-8,905 PL with an average 8,413 PL during second year (Fig. 3b). The same trend with lower PL production was recorded in medium 2 (brine solution) which ranged from 6,498-7,825 with an average of 7,246.3 PL during first year and lower PL production 6,000-7,096 with an average 6,574.3 PL in medium 2 during second year. Similarly, the overall average 6,910.3 PL Production and 22.85 PL/L was recorded for 2 years in medium 2 which was lower than that of medium 1 (Fig. 3d). First few PL were observed on day 19 and the trials concluded on day 40. Average PL production exhibited peak on day 30-31. While comparing the average PL production of two years in the two media, better result was obtained in natural seawater (medium 1) as compared to brine solution.

Medium 1 first year

8726   9009

8946
☐ T1 ☐ T2 ☐ T3 (a)

Medium 1 second year

7744   8905

8590
☐ T1 ☐ T2 ☐ T3 (b)

Medium 2 first year

6498   7825

7419
☐ T1 ☐ T2 ☐ T3 (c)

Medium 2 second year

6000   7096

6627
☐ T1 ☐ T2 ☐ T3 (d)

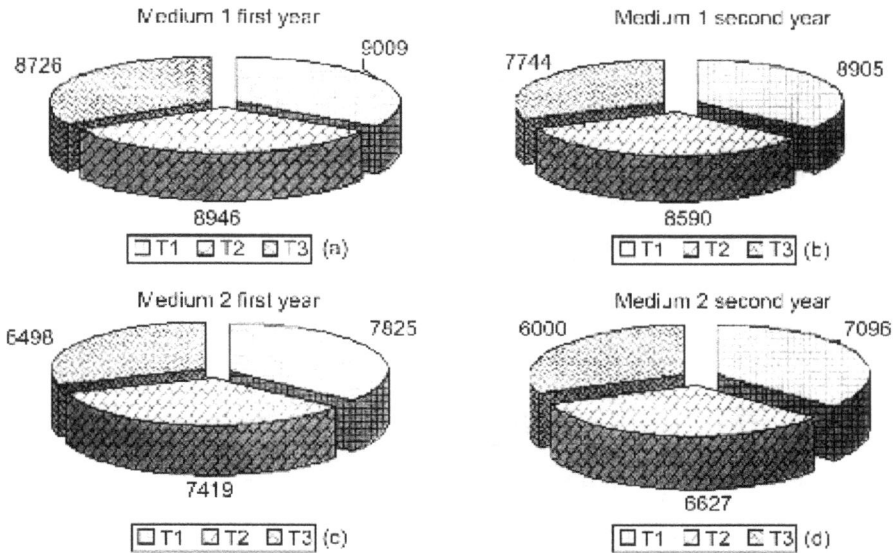

FIG. 3. (a-d) Post-larval production recorded in three trails in medium 1 and 2 during 2003 and 2004.

Larval behaviour: Freshwater prawn larvae eat continuously but they do not actively search for food. In the present study, the larvae were fed with *Artimia* nauplii alone up to day 7, thereafter food was provided in combination with it. It was observed that the acceptance of *Artemia* nauplii was better than that of egg custard and mussel meat. All the 11 larval stages showed propensity to feed *Artemia* nauplii, egg custard and mussel meat which were most acceptable feed provided on day 8 onwards. Feed quality, quantity and feed schedules are very important to achieve successful post-larval production (Sounarapandian and Kannupand, 2000). The exact quantity of food required at each meal depends upon the utilization of food by the larvae which need to be judged visually (Aquacop, 1977; New and Singholka, 1985). Suitable size of food particles for different stages of larvae of *M. malcolmsonii* has been suggested by Kanaujia and Mohanty (1992) and Kanaujia (1998, 1999). In this experiment, the appropriate size of food particles and schedule of feeding (4 times/day) proved useful for better production of post-larvae which corroborates with the observations of Kanaujia and Mohanty (1992) and Joshi and Raje (1992).

Water quality: Water quality in a hatchery system is most important factor which plays important role in the growth, metamorphosis and survival of larvae (Boyd, 1990; FAO, 2002b). The various water quality parameters such as salinity, temperature, pH, total hardness, total alkalinity,dissolved oxygen, ammoniacal nitrogen *etc.* in the present study werewithin the desired range for effective larval rearing.

Water temperature: Water temperature regulates the metabolism and growth of various larval stages of prawn and the favourable temperature range being 28-31°C for optimum larval growth and development of *M. rosenbergii* (FAO, 2002b; New and Valenti, 2009) and *M. malcolmsonii* (Kanaujia and Mohanty, 1992; Prasad and Kanaujia, 2006). Since the present study was carried out under similar conditions, variations in ambient temperature of the rearing medium was minimum and insignificant among the treatments. Water temperature was slightly higher in natural seawater during the second year as compared to first year. Temperature above 35°C and less than 24°C retard growth and mortality in the larvae of *M. rosenbergii* (New and Singholka, 1985; Diaz and Ohno, 1986; FAO, 2002b; New and Valenti, 2009).

Salinity: The salinity is mainly due to carbonates and sulphate of sodium and potassium (Boyd, 1990). As salinity affects the rate of growth, the larvae need a suitable range for optimum development (Katre, 1977; Boyd, 1990; Kanaujia and Mohanty, 1992; Mitra, 2001; Kanaujia et al., 2001; New and Valenti, 2009). Kanaujia *et al.* (2001, 2005) reported and optimal salinity range of 12-16 ppt for larval growth and development of *M. gangeticum*. The salinity range during larval culture of *M. rosenbergii* varies from 10-18 ppt (Brock, 1993; Thongrod et al., 2007). A sudden change in ambient salinity over 10 ppt may be fatal to larvae as they fail to acclimatize under altered environmental condition. Salinity range between 12-20 ppt was been considered optimum for completion of all larval stages of *M. rosenbergii* (New and Singholka, considered optimum for completion of all larval stages of *M. rosebergii* (New and Singholka, 1985 FAO, 2002; Hangsapreurke *et al.*, 2008) and *M. malcolmsonii* (Kanaujia and Mohanty, 1992; Kanaujia et al., 2005). In the present study, the salinity range 10-16 ppt in the two media during 12 trials was found to be optimal for larval growth and development of *M. gangeticum*.

pH: pH is an important factor in determining the productivity in aquaculture (Boyd, 1990). It measures the hydrogen ion concentration in the water and therefore serves as indicator of acidity and alkalinity (Mohanty, 2003). Most of the biological parameters of aquatic water bodies are influenced by pH. Generally, pH in water is affected by the concentration of free carbon dioxide, bicarbonate and carbonate ions which is greatly influenced by the presence of residual feed, larval metabolites and moulting shells (Kanaujia and Mohanty, 1992). pH of the rearing medium has been recognized as important parameters during larval rearing of *Macrobrachium* spp. The total dissolved ammonia affects the survival particularly during moulting when pH of the water is high. Therefore, to avoid possible toxicity of ammonia in prawn hatchery, Kanaujia and Mohanty (1992) suggested for maintaining water pH within the range of 7.5-8.5. New and Singholka (1985) reported a suitable range pH between 7.5 and 8.5 during larval rearing of *M.rosenbergii*. In the present study, pH was maintained in range of 7.5-7.7 in the two media of larval rearing in all the 12 trials which was slightly lower as compared to the earlier reports. However, the initial higher pH observed in trials did not continue for longer period due to application of calcium sulphate and calium hydrogen phosphate in the bio-filter tanks.

Dissolved oxygen (DO): The dissolved oxygen in water medium directly affects the growth and survival of the larvae. Due to brackishwater environment, DO level was maintained over 4 mg/*l* during the cycle in the test media in spite of continuous aeration. DO value was recorded lower (4.2 mg/*l*) during first year. The mean values of DO differed significantly between first and second year. In the present study in four mediums of 12 trials, the oxygen varied between 4.0-4.9 mg/*l* which is found a very narrow range. It was highest in medium 1 and lowest in medium 2. The wide variations in dissolved oxygen during larvae culture of *M. malcolmsonii* were reported by Mohapatra (2001), which was a variation in climatic temperature and disruption of power failure during hatchery operation. The optimum levels of dissolved oxygen in the larval rearing medium of prawns commonly maintained through water aeration with the use of aerators/air compressor/air blower (Brock, 1993; Chowdhury *et al.*, 1993; Hangsapreurke *et al.*, 2008). Dissolved oxygen is important not only for respiration but also for maintenance of most favourable chemical and hygienic and environmental conditions of the larval rearing medium (FAO, 2002b; New and Valenti, 2009). However, in the present study the mean value recorded, was not significantly different in the trials.

Total hardness: In the test media, significant variations (p<0.05) in the total hardness were recorded during first and second year of operation. However, higher values have been well registered in second year than first year. Freshwater prawns as well as most of the crustaceans require high calcium concentrations for the enzymatic processes involed in moulting and there is also a relationship between magnesium and neural muscular energy transmission (Hangsapreurke *et al.*, 2008). Total hardness affects the growth of the larvae and mineralization of carapace (FAO, 202b; New and Valenti, 2009). Thus, it can be concluded that among the two mediums under present larval rearing trials, medium 1 showed better efficiency in terms of growth, reproduction, larval development, salinity requirement, duration of larval cycle etc. than those of medium 2, under the similar environmental conditions.

Hardness of water caused due to the carbonate, bicharbonate, chloride and sulphate ions in association with calcium and magnesium. In the present study, the total hardness in two mediums ranges from 2,230-2,292 mg/l during two years. The optimum total hardness level needed for larval metamorphosis was reported within the range of 3,800- 5,200 mg/l in *M. malcolmsonii* (Kanauji and Mohanty, 1992). Mohapatra (2001) recorded total hardness level within 2,020-2,220 mg/l as calcium carbonate in the comparative larval rearing study of *M. malcolmsonii* and *M.rosenbergii* and found within the desired level. However, it differed form those of Kanaujia and Mohanty (1992), they had recorded higher-level of hardness 3,800-5,200 mg/l in *M.malcolmsonii*. Brown *et al.* (1991) found that the growth of *M.rosenbergii* was maxium at hardness levels below 53 mg/1 as $CaCO_3$ and survival was at higher rates.

Total alkalinity: The total alkalinity of water is mainly caused by the contents of Ca, Mg, Na, K, NH_4 and Fe combined either with carbonates, bicarbonates or occasionally by hydroxide (Jhingran, 2003; New and Valenti, 2009). In the present observations, total alkalinity ranged from 79.2-91.7-mg/l in the two media. Total alkalinity denotes the quantity of acid consuming constituents present in the water. In natural water, bicarbonates and carbonates are the main alkaline sources which determine pH of water. Water with low alkalinity exhibits low buffering action leading to wide range of fluctuations in pH value. High alkalinity increases pH and causes larval mortality. Alkalinity ranging from 50-100 mg/1 have been reported as desirable level for *M. rosenbergii* larvae (Chandraprakash and Reddy, 1993).

Ammoniacal nitrogen: Ammonia exists in water in two forms, namely un-ionized ammonia (NH_3^+) and ammonium ion (NH_4^+) (New, 2002; Mohanty, 2003). Ammonical nitrogen in water medium is an important factor which directly influences the life of the oraganisms present in aquatic ecosystem. In the present study, maximum value of ammonical nitrogen found during the second year was significantly higher than first year. Its was observed that there was no accumulation of ammonia in the test media during both the years. The rise in pH and temperature of water increases the concentration of un-ionized fraction of ammonia. Compared to ammonium ions, the un-ionized ammonia is highly toxic to the larvae. The total ammonia (NH_3^+ and NH_4^+) less than 0.12 ppm in the rearing medium considered 'safe' for prawn larvae (Kanaujia and Mohanty, 1992; Kanaujia, 1998, 1999). In present study, it ranged from 0.072-0.137 mg/ l in two of 12 trials which was much below the 'safe level'. An initial accumulation of excretory ammonia was found in all the trials of larval rearing of *M. gangeticum*. Interestingly, the average ammonia content was significantly lower in some of the trials which might be due to the ineffective nitrification of chelated ammonia in the rearing trials in the presence of Na-EDTA which was applied @ 5 mg/ l every alternate day in larval rearing tanks.

   PL production: Commercially important larger *Macrobrachium* species are mostly estuary-bound and need brackishwater for completion of their larval phase (Rajyalaxmi, 1980). Some of the species like *M. malcolmsonii* and *M. gangeticum* migrate long distance from estuary to inland region where climatic conditions are suitable for their growth. Collection and transportation of their seed form natural resources is the only option to use for culture which is uncertain and irregular. Some workers have tried seed production of *Macrobrachium* species using brine (salt pans) diluted with well or surface water and artificial seawater (Tunsutapanich, 1980; Kanaujia and Mohanty, 192; Chowdhury *et al.*, 1993; Hangsapreurke *et al.*, 2008). A suitable and effective larval rearing medium is the key to success in large-scale prawn seed production (New, 2002; New and Velenti, 2009. All the 12 trials conducted in the two media in the present experiment resulted in successful metamorphosis to the post-larvae of *M. gangeticum*. Although the appearance of first post-larva was noticed earlier in medium 1 in trial 1 during first year, the total number of post-larvae produced was highest (9,009) followed by medium 2. While comparing the average PL production in 12 trials during two years of the two media, the natural seawater (medium 1) gave better result followed by brine solution (medium 2).

CONCLUSIONS

In present study, the trials carried out in natural sea water and concentrated brine solution resulted better production of the post larvae. Successful hatchery operation for large-scale seed production of freshwater prawn requires suitable and effective larval rearing medium. Occurrence of first few post-larvae and completion of production cycle was observed earlier than those of *M. rosenbergii* and *M. malcolmsonii*. The provision for larval food, effectiveness of rearing medium and technique may also have a bearing on the growth and survival of post-larvae. Further, the post-larval production 4,748- 9,009 @ 15-30 PL/L along with better survival rate (15.82-30.03 %) and shorter larval duration (40 days) indicated the suitability of both the media for hatchery operation. Closing of larval cycle of *M.gangeticum* in brine solution indicate to the inclusion of one more *Macrobrachium* species in freshwater prawn farming. Further investigations are also needed on husbandry, management and composition of feed ingredients for various larval stages so that survival rate, salinity requirement and duration of larval cycle be reduced further to produce post-larvae on mass-scale aquaculture of *M. gangeticum* in inland areas far away from the coastal regions.

ACKNOWLEDGEMENTS

The senior author (PM) wishes to express her gratitude to the Director, Central Institute of Freshwater Aquaculture, Bhubaneshwar for permission to carry out the research work and to the Head, Department of Zoology, Utkal University, Bhubaneshwar for Providing necessary laboratory facilities.

REFERENCES

APHA, (1985). *Standrad Methods for the Examination of Water and Wasterwater. 16th Edition.* American Public Health Association, Washington D. C., p. 1268.

Aquacop, (1977). *Macrobrachium rosenbergii* culture in Polynesia, progress in developing a mass intensive larval rearing technique in clear water. *Proc. World Maricult. Soc.,* 8: 311-326.

Brown, J.H., J.F. Wickins and Maclean, M.H., (1991). The effect of water hardness on growth and carapace mineralization of juveniles of freshwater prawn *M. rosenbergii. Aquaculture,* 95: 329-345.

Brock, J.A., (1993). A synopsis of pathology, diseases and production problems of broodstock rearing of the giant prawn, *M. rosenbergii* (de man). A critical factor for oviposition and larval production. In: *International Symposium on Freshwater Prawns-2003* (August 21-23, 2003). Kerala Agricultural University, College of Fisheries, Cochin, pp. 145-147.

Boyd, C. E., (1990). *Water Quality in Ponds for aquaculture.* Birmingham Publishing Company, Auburn University, Alabama(USA). p. 482.

Chandraprakash and Reddy, A.K., (1993). Water quality management in giant freshwater prawn hatchery. In: *Short-term Training Programme on Hatchery Management of Freshwater Giant Prawn in Artificial Sea Water.* Central Institute of Fisheries Eductation, Bombay, pp. 44-47.

Chowdhury, R., Bhattacharjee, H. and Angell, C., (1993). *A Manual for Operating a Small-Scale Recirculation Freshwater Prawn Hatchery.* FAO Corporate Document Repository. FAO-Bay of Bengal Programme, Chennai.

Diaz, G. G. and Ohno, A., (1986). Possible significance of rearing conditions of ovigerous *Macrobrachium rosenbergii* (de Man). In: *Processdings of the First Asian Fisheries Forum* (May 26-31, 1986) (J.L. Maclean, L.B. Dizon and L.V. Hosilles, eds.). Asian Fisheries Society, Manila, Philippines, pp. 45-48.

FAO, (2002a). *Fishery Statistic: Aquaculture Production (2000): FAO Year Book.* Food & Agriculture Organization, Rome, Italy, p. 180.

FAO, (2002b). *Farming Freshwater Prawns: A manual for Culture of Giant River Prawn (Macrobrachium rosenbergii).* FAO Fisheries Technical Paper No. 428.

Hangsapreurke, K., Thamrongnawasawat, T., Powtongsook, S., Tabthipwon, Lumubol, P. and Pratoomchat, B., (2008). Embryonic development, hatching, mineral composition, and survival of *Macrobrachium rosenbergii* (de Man) reared in artificial searwater in closed recirulating water system at different levels of salinity. *Mj. Int. J. Sci. Tech.,* 2:471-482.

Jhingran, V. G., (2003). *Fish and Fisheries of India.* Hindustan Publishing Corporation, New Delhi.

Joshi, V. P. and Raje, P. C., (1993). Packaging and transportation trials with the seedlings of the gaint freshwater prawn, *Macrobrachium rosenbergii.* In: *Proceedings of the National Seminar on Aquaculture Development in India: Problems and Prospects* (November 27-29, 1990) (P. Natarajan and V. Jayaprakash, eds.). Departments of Aquatic Sciences, University of Kerala, Thiruvananthapuram, pp. 21-26.

Kanaujia, D.R., (1998). Emerging technologies in seed production of *Macrobrachium malcolmsonii* (H. Milne. Edwards). In: *Processings of the Current and Emerging Trends in Aquaculture* (P.C. Thomas, ed.). Daya Publishing House, New Delhi, pp. 148-159.

Kanaujia, D.R., (1999). Seed production of Indian river prawn *Macrobrachium malcolmsonii* (H. Milne Edwards). In: *Aquaculture* (B.C. Mahapatra, P.G. Ingole and G.M. Bharad, eds.). Dr. Panjabrao Deshmukh Krishi Vidyapeeth, Akola, (Maharashtra), pp. 212-227.

Kanaujia, D.R., (2006). Freshwater prawn breeding and culture In: *Handbook of Fisheries and Aquaculture* (S. Ayyappan, J.K. Jena, A. Gopalakrishnan and A.K. Pandey, eds.). Directorate of Information and Publication in Agriculture (ICAR), New Delhi, pp. 293-306

Kanaujia, D. R., Mohanty, A.N. (1992). Breeding and large-scale seed production of the Indian river prawn *Macrobrachium malcolmsonii* (H. Milne Edwards) J. Aqua., 2:7-16.

Kanaujia, D.R. and Mohanty, A. N., (2001). Effect of salinity on the survival and growth of the juveniles of Indian river prawn, *Macrobrachium malcolmsonii* (H. Milne Edwards). J. *Adv. Zool.,* 22:31- 36.

Kanaujia, D.R. and Mohanty, A.N. and Soni, S., (2001). Breakthrough in seed production of Ganga river prawn, *Macrobrachium gangeticum* (Bate, 1868). A milestone in aquafarming Fishing Chimes, 21 (1): 28-30.

Kanaujia, D.R., Mohanty, A.N. and Soni, S., (2002). String Shell technique for effective harvesting of the post-larvae of Indian river prawn under hatchery condition. *Fishing Chimes,* 22(1): 117-119.

Kanaujia, D.R., Mohanty A.N. and Mitra, G. and Prasad, S., (2005) Breeding and Seed production of Ganga river prawn, *Machobrachium gangeticum* (Bate) under captive condition. *Asian Fish. Sci.,* 18: 371-389.

Kanaujia, D.R., Mohanty A.N. and Mohanty, U.L., (2007). Seed production of the Ganetic prawn, *Macrobrachium gangeticum* (Bate) in India. In: *Freshwater Prawns: Adances in Biology, Aquaculture and Marketing* (C.M. Nair, D.D. Nambudiri, S. Jose, T.M. Sankar, K.V. Jayachandran and K.R. Salin, eds.). Allied Publishers, New Delhi, pp. 310-314.

Karte, S., 1977. Yolk utilization in the freshwater prawn, *Macrobrachium rosenbergii. J. Amin Morphol. Physiol.,* 24: 13-20.

Ling, S.W., (1969). Methods of rearing and culturing *Macrobrachium rosenbergii* (de Man). *FAO Fisheries Report,* 57(3): 607-619.

MPEDA, (2001). *An Overeiw: Marine Products Export Development Authority*, India. Ministry of Commerce & Industry, Government of India. Cochin, p. 47.

MPEDA, (2004). *An Overview; Marine Products Exports Development Authority, India.* Ministry of Commerce & Industry, Government of India, Cochin, p. 47.

Mitra, G., (2001). Effect of progressive increase in the medium salinity on the growth and post larvae production in Indian river prawn, *Macrobrachium malcolmsonii* (H. Milne Edwards). M.F.Sc. Thesis. Department of Applied Aquaculture, Barkatullah University, Bhopal, p. 120.

Mohanty, M., (2003). Studies on the larval growth, survival and post larval production of gaint freshwater prawn, *Macrobrachium rosenbergii* (de Man) in relation to water qualities. Ph. D. Thesis. Utkal University, Bhubaneshwar, p. 148.

Mohapatra, J., (2001). Studies on the comparative breeding and larval biology of Indian river prawn *Macrobrachium malcolmsonii* (H. Milne Edwards) and giant freshwater prawn, *Macrobrachium rosenbergii* (de Man). Ph.D. Thesis, Utkal University, Bhubaneshwar, p. 280.

New, M.B., (2002). *Farming Freshwater Prawns: A Manual for the Culture of the Giant River Prawn (Macrobrachium rosenbergii).* FAO Fisheries Technical Paper No. 428. FAO, Rome, Italy, p. (2012).

New, M.B. and Singholka, S., (1985). *Freshwater Prawn Farming: A Manual for the Culture of Macrobrachium rosenbergii.* FAO, Fisheries Technical Paper, pp. 118-222.

New, M.B. and Valenti. W.C., (2009). *Freshwater Prawn Culture: the Farming of Macrobrachium rosenbergii.* Blackwell Science, Oxford, England, p. 512.

Prasad, S. and Kanaujia, D.R., (2006). Availabiltiy and breeding behavior of Ganga river prawn, *Macrobrachium gangeticum* (Bate) and *M. malcolmsonii* (H. Milne Edwards). *Asian Fish. Sci.,* 19: 377-388

Rajyalaxmi, T., (1980). Comparative study of the biology of the freshwater prawn *Macrobrachium malcolmsonii* of Godavari and Hooghly river system. *Proc. Indian Natl. Sci. Acad. (Anim. Sci.),* 46B: 72-86.

Sounarapandian, P. and Kannupand, T., (2000). Effect of feed of *Macrobrachium malcolmsonii* reared in synthetic brackishwater. Indian J. Exp. Biol., 38: 287-289.

Thongroad, S., Sauggontangit, T., Tamtin, M., Chaihul, S.L., Tunsutapanich, A. and Boonyaratpalin, M., (2007). Seed production of giant freshwater prawn, *Macrobrachium rosenbergii, in earthen ponds.* In: *Freshwater Prawns: Advances in Biology, Aquaculture and Marketing* (C.M. Nair, D.D. Nambudiri, S. Jose. T.M. Sankar, K.V. Jayachandran, and K.R. Salin, eds.). Allied Publishers, New Delhi, pp. 344-348.

Tunsutapanich, A., (1980). The use of rock salt brine and salt stock solution for larval culture of *Macrobrachium rosenbergii.* UNDP/FAO Programme for the Expansion of Freshwater Prawn Farming Working Paper. FAO, Rome. THA /75/008/80/WP/18, pp. 1-8.

□□□

3

INLAND FISHERIES IN RAINFED AREAS OF JAMMU, J&K STATE, INDIA

B.L. Kaul and Rajesh Dogra

ABSTRACT

The paper deals with the Inland Fisheries potential in the ponds and tanks of rainfed areas popularly called Kandibelt of Jammu. The Kandibelt has a varied topography produced by Sivalik hills. It is barren, broken undulating, highly eroded and dry landscape dotted with numerous ponds, tanks and a couple of lakes. Because of the undulating type of terrain, the runoff is generally rapid. Traditionally ponds were dug and maintained by village elders for agriculture , washing, drinking and for fisheries prior to 1947. However, many of these water bodies became degraded and were lost for use after the independence. Efforts are now on to rejuvenate and reuse old ponds and create new ones for the purpose of fisheries.

INTRODUCTION

In India approximately 14.49 million people get their livelihood from fisheries sectors. Thus this sector contributes significantly to the national economy. In India fisheries sector has shown a sharp growth with an annual fish production of 7.6 million ton during 2008-09 from just 0.75 million ton in 1950-51. The contribution of Inland fisheries sector was 4.6 million ton in 2008-09 compared to 0.218 million ton during 1950-51. Despite this sharp growth in the sector there still is need for technological and policy intervention to sustain this growth rate. There is indeed need for innovative, locally adaptive technologies along with well defined extension approach to sustain the growth in inland fisheries sector (Mishra 2012). Needless to say that in order to improve inland fisheries substantial expertise and investments to improve water quality and quantity will also be required.

An important segment of inland fisheries in India lies in the rainfed areas. A considerable population of traditional fisherman from rainfed regions in Maharashtra, West Bengal, Assam, Karnataka, Andhra, Tamil Nadu, Odisha, Madhya Pradesh, Bihar, Jharkhand, Chhatisgarh, U.P., Gujarat, Himachal Pradesh and Jammu and Kashmir are dependent on fisheries from rivers and traditional ponds and tanks. Promotion of tank and pond fisheries with the help of UNDP and other agercies of G.O.I. has helped ordinary people and marginalized sections of the society in a big way in overcoming poverty and hunger. In Maharashtra, for example, there are thirteen thousand traditional tanks in just two districts of Bhandara and Gadchiroli and these form backbone of economy of fishermen, nomadic tribes and scheduled tribes in the region (Dandekar 2012).

Similarly in rainfed areas of West Bengal which have undulated topography and rock soil with low water retention small and marginal farmers are encouraged to excavate new ponds and re-excavate old ones for fish culture (Das 2012). In Karnataka there are 600 community based tanks in rainfed areas providing source of livihood to 4850 persons of whom 75% are landless and 25% are marginal farmers. In Andhra Pradesh also community based tank projects have helped the poor to improve their economic standards through community participation. Microfinancing has been used in agriculture as a helping tool in proverty reduction.

Rice fish farming and village tank aquaculture in rainfed areas of Odisha and promotion of inland fisheries in rainfed areas of Tamil Nadu are other examples of pisciculture helping poor farmers in increasing incomes and meeting their nutritional requirements. In Jammu region of Jammu and Kashmir state Kandi areas of Jammu, Samba and Kathua Districts, community/village ponds are also helping the landless and marginal farmers in improving their livelihood (Kaul 2012). Under Rashtriya Krishi Vikas Yojna and PM's employment package a number of ponds have been constructed and this has helped many unemployed in fish culture.

STUDY AREA

The Kandi belt of Jammu region lies between the longitude of 74°-45' East and longitude of 32°-22' North to 32°-55' North except in the western portion where it lies between latitude of 32°-50' North to 33° N. The belt includes five districts namely Jammu, Samba, Kathua Reasi and Udhampur. The present study is restricted to the first three disducts and excludes Reasi and Udhampur districts. The Kandi area is about 200 km long and 10 to 20 kms wide between the river Ravi on the east and Manawar Tawi river on the west. The southern boundary runs beyond Jammu along the Ranbir canal to Akhnoor area and then along the Pratap canal to the line of control on the Manawar Tawi. The Siwalik range rising to about 740 meters height above the mean sea level constitutes the northern border. The lower Southern border is about 296 meters height just where the foot hills merge into the plains. The total area of the Kandi region is about 4,89,266 hectares, out of which only 1,16,046 hectares (23.76%) is cultivated (Raina 1992).

TOPOGRAPHY AND CLIMATE

The Kandi belt has a varied topography producted by Siwalik Hills. Because of undulating and broken type of terrain, the run off is generally rapid. The Kandi belt areas have traditionally been water scarcity area for a very long time past. In the beginning of twentieth century forests were cut off to supply fuel-wood for brick-kilns. The denudation is so extensive that even drinking water in most of the springs has dried.

The region has broadly two climate seasons. The hot season lasts from April to October for 7 months with maximum mean temperature of 34.5^oC and minimum mean temperature of 24.3^oC. From November to March is the cold season for 5 months with mean maximum temperature averaging 22^oC and the mean minimum temperature of 11.5^oC. The annual rainfall in this region ranges between 900-1000 mm. 75 percent of rainfall is received during the monsoons. The distribution of rainfall in the region is highly erratic. It frequently results in droughts of moderate to severe intensity. June and September months generally exhibit severe drought conditions.

SOILS

The soils of the Kandi belt are alluvial soils ranging from sandy loam to loamy sand having different depths and slopes. In some areas the soils are very shallow. Stony and deep soils are also present in patches. Loamy sand is the predominant texture on the surface soils, but the clay contents increase with the depth of soil. (Raina and Pikhan 1992).

PHYSIOGRAPHY

On the basis of physiographic features, climate conditions and watersheds of the Ravi and the Chenab which flow in the Kandi belt, it is divided into four zones:

1. Lower Ravi Kandi basin;
2. Upper Ravi Kandi basin;
3. Lower Chenab Kandi basin; and
4. Upper Chenab Kandi basin.

OBSERVATIONS AND DISCUSSION

In Jammu Kandi area (rainfed area of Jammu Division which covers parts of the districts of Kathua, Samba, Reasi, Jammu and Udhampur) it is believed that in 1947 there were nearly 800 ponds of different sizes. Since the area lies in the Sivaliks most of the rainwater is lost in surface run off and there is scarcity of water for most part of the year. However, the wise elders of the area had created ponds which used to store water for use during the rainy season. This water was used for agriculture, drinking, washing as well as for fishery.

Unfortunately, after independence these ponds were neglected, filled up, polluted and in same places grabbed by influential people for their personal use. Urbanization of rural areas also played its role in degrading these ponds. At present, there are only 282 ponds left in the study area (Table 1). Nearly 200 of these ponds are stocked annually by the State Fisheries Department with seed of culturable carps for the benefit of the people resulting in fish harvest of 253125 kg and income of 171.57 lacs during 2011-12 (Table 1). A breakup of the details of types of ponds, their number, water area (in ha) number of ponds stocked, area of ponds, quality of seed stocked, fish harvest (in kgs) and the resultant income is given together for the three districts under study of the Kandi belt in Table 1 and separately for Jammu district (Table 2), Samba district (Table 3) and Kathua district (Table 4).

Despite efforts of the fisheries department production is poor for a number of reasons such as:

1. Urbanization of villages has resulted in construction of drains which carry sewage into these ponds and pollute their water.
2. There is poor management because Village Panchayats do not control them at most places.
3. Because of the lack of proper local management, everybody is free to indulge in theft and steal fish. The Government cannot afford to post guards at each and every pond.
4. There is no proper arrangement to fertilize the ponds and the only food available to the fish are the naturally growing weeds and plankton because of which fish size remains small.

Table 1: Details of ponds in Kandi areas (District Jammu, Kathua, Samba)

S.No.	Type of ponds	Total Nos. of ponds available 1	Water area (in ha) 2	No. of ponds stocked 3	Area of ponds Stocked (in ha) 4	Quantity of seed stocked 5	Fish havest (in kg) 6	Economic activity (in Rs.) 7
1.	Community/ Village ponds	237	73.75	174	66.95	1104000	222425	15123500
2.	PM's package ponds constructed under PM's Employment package	4	0.40	3	0.30	6000	2750	192500
3.	Ponds constructed under Rashtriya Krishi Vikas Yojana (RKVY)	6	1	6	1	16000	3400	238000
4.	Private ponds	17	2.70	17	2.70	10000	17550	1113000
5.	Other schemes	18	2.00	18	2.00	102000	7000	490000
	G. Total	282	79.47	218	72.57	1238000	253125	17157000

(171.57 lacs)

Table 2: Details of ponds in Kandi areas (District Jammu)

S.No.	Type of ponds	Total Nos. of ponds available 1	Water area (in ha) 2	No. of ponds stocked 3	Area of ponds Stocked (in ha) 4	Quantity of seed stocked 5	Fish havest (in kg) 6	Economic activity (in Rs.) 7
1.	Community/ Village ponds	91	37.2	91	37.2	576000	130200	9114000
2.	PM's package ponds constructed under PM's Employment package	3	0.3	2	0.2	4000	2400	168000
3.	Ponds constructed under Rashtriya Krishi Vikas Yojana (RKVY)	2	0.4	2	0.4	8000	1st year of Stocking 2400 (Expected)	168000
4.	Private ponds	7	0.6	7	0.6	10000	6000	420000
5.	Other schemes	18	2	18	2	102000	7000	490000
	G. Total	122	40.5	120	4.4	700000	148000	10360000

(103.00 lacs)

Table 3: Details of ponds in Kandi areas (District Samba)

S.No.	Type of ponds	Total Nos. of ponds available 1	Water area (in ha) 2	No. of ponds stocked 3	Area of ponds Stocked (in ha) 4	Quantity of seed stocked 5	Fish havest (in kg) 6	Economic activity (in Rs.) 7
1.	Community/ Village ponds	67	17.85	44	17.85	282000	44625	2677500
2.	PM's package ponds constructed under PM's Employment package							
3.	Ponds constructed under Rashtriya Krishi Vikas Yojana (RKVY)	2	0.20	2	0.20	4000	Stock in July 400-500gms	
4.	Private ponds	10	2.10	10	2.10	0	11550	693000
5.	Other schemes							
	G. Total	79	20.15	56	20.15	286000	56175	3370500

(33.705 lacs)

Table 4: Details of ponds in Kandi areas (District Kathua)

S.No.	Type of ponds	Total Nos. of ponds available 1	Water area (in ha) 2	No. of ponds stocked 3	Area of ponds Stocked (in ha) 4	Quantity of seed stocked 5	Fish havest (in kg) 6	Economic activity (in Rs.) 7
1.	Community/ Village ponds	79	18.7	39	11.9	246000	47600	3332000
2.	PM's package ponds constructed under PM's Employment package	1	0.1	1	0.1	2000	350	24500
3.	Ponds constructed under Rashtriya Krishi Vikas Yojana (RKVY)							
4.	Private ponds	2	0.02	2	0.02	4000	1000	70000
5.	Other schemes							
	G. Total	82	18.82	42	12.02	252000	48950	3426500

(34.26 lacs)

5. In some places the locals do throw vegetable waste, rice bran and gobar in the ponds which helps weed growth and fish size is comparatively larger.

6. In some villages there is a temple by the side of a pond. The priests let the Fisheries Department to stock the ponds but do not allow fishing, due to religious sentiment.

7. The soil in Kandi area is porous and water retaining capacity of ponds is poor. If pucca ponds are built or existing ponds are lined with HDPE/LDPE the situation can improve but it will involve large sums of money. However, some ponds have been constructed by the Fisheries Department under P.M'S package and RKVY.

8. Efforts need to be made to improve watershed management so that water can be stored in existing as well in newly created ponds.

The increase of fish production in small water bodies can be realized basically by a balanced approach of optimizing the utilization of the whole ecosystem, manipulation of stock and species, manouvring habitats favorable to target fish species, enriching nutrients level for growth and survival. All of these are the basis of enchancement interventions converging on combined attributes of aquaculture for increasing production. Most of the enhancement technologies are framed with detailed consideration on physical (morphometric, edaphic and hydrologic) and biological (food web, life history, species intervention, carrying capacity) factors to achieve closer to natural limit to production potential.

Carp polyculture is a versatile form of culture system that easily integrates with forestry, agriculture, horticulture and livestock based system with high degree of complimentarity in terms of resource use and production benefits. In Jammu Kandi belt there is immense scope for adoption of agroforestry system such as forestry-agronomy-fishery and forestry-agronomy-animal husbandry-fishery (Kaul 1996). Chatterjee (2012) has opined that integrating fish raising with ducks, pigs, goats and chickens etc in rainfed areas has proved useful in West Bengal and Bangladesh.

Aquaculture provides food security to hungry people. Innovations in aquaculture projects supported by World Bank in Assam, Karnataka and Andhra, have helped ordinary people and women in marginalized sections of society there. The poor can also improve their economic standards through community participation by addressing common property use of natural resources. These innovative sustainable projects have shown that the aquaculture through microfinance can be the best tool for achieving reduction of poverty, gender participation and production of employment in Karnataka and Andhra Pradesh (Upare 2012) and in Tamil Nadu (Madan Mohan 2012).

In Jammu and Kashmir the Government has been encouraging local people in the Kandi belt to use existing ponds and to create new ponds for fishery. For this purpose financial assistance to the extent of Rs. 1.5 lacs is provided under various schemes for fish culture. Many units have come up and thus generated employment to the unemployed poor.

According to Biswas (2012) rice-fish farming is one of the best farming options for the less productive rainfed ecosystems. It is practiced in many rainfed regions in North-East, Odisha, West Bengal and Andhra Pradesh. Rice-fish farming in rainfed areas provides the much needed synergy among myriad components like rice, fish, prawns and other fauna thus promoting and harnessing biodiversity. Their interaction results in enrichment of soil nutrient status, better crop nutrition and biocontrol of pests and weeds. Not only does it result in fishery development in rainfed conditions, it has been proved to also increase rice yields by up to 20 percent besides reduction in application of chemical fertilizers and pesticides. Poor and marginal farmers need to be encouraged to adopt such simple cost effective methods. Although not tried in Jammu Kandi belt, yet there seems enough scope for trying it.

FIG. 1. Fish seed stocking in community pond at Mera Mandrian, Akhnoor Dist. Jammu

FIG. 2. Jihdrah, Block Dansal Dist. Jammu

FIG. 3.   Fish seed stocking in community pond at Pallanwala, Block Khour Dist. Jammu

FIG. 4.   Fish seed stocking in community pond constructed under MGNAREGA at Upper Kathar, Block Dansal Dist. Jammu

FIG. 5.   Stocking of community pond at Badla, Block Chagwal Dist. Samba 2012-2013

FIG. 6. Stocking of community pond at Papad Block Sarba, Dist. Samba 2012-2013

Fig. 7. Stocking of community pond at Patti Block Vijaypur Dist. Samba 2012-2013

FIG. 8. Stocking of community pond at Jaloor Block Samba, Dist Samba 2012-2013

FIG. 9.   Stocking of community pond at Bhadhu I Billawar, Dist Kathua 2012-2013

FIG. 10.   Stocking of community pond at Bhadhu II Billawar, Dist Kathua 2012-2013

Fig. 11.   Stocking of community pond at Barote Billawar, Dist. Kathua 2012-2013

Fig. 12. Stocking of community pond at Charn Morian, Dist. Kathua 2012-2013

ACKNOWLEDGEMENT

The senior author is thankful to Mr. Gopi Ghosh of F.A.O. in India and Bhutan for permission to quote from the responses received in UNDP Solution Exchange on the query of Mr. Neelkanth Mishra: Promoting Inlan fishery (ftp:// ftp.solution.exchange.net.in / public / food / resource / res 14121101. docx)

REFERENCES

Biswas Anibrata, Sir Dorabji Trust (SDTT)-SRI Programme Bhubaneshwar, Odisha (2012) UNDP solution exchange, New Delhi.

Chatterjee A.S. DRCSC, Kolkatta, W.B. (2012) UNDP Solution Exchange, New Delhi.

Dandekar Parineeta, South Asia Network on Dams, Rivers and people, New Delhi, (2012) UNDP Solution Exchange New Delhi.

Das Anushuman (2012), Development, Research communication and services centre (DRCSC) Kolkata, W.B. (2012) UNDP Solution Exchange, New Delhi.

Kaul B.L, Food Production Potential of Jammu Kandi Belt (J&K) in Food security and Panchyati Raj (Pradeep Chaturvedi Ed) (1996) Concept Publishing House, New Delhi.

Kaul B.L., Society for Popularization of Science, Jammu, (2012) UNDP Solution Exchange New Delhi.

Madan Mohan DHAN Foundation, Madurai Tamil Naidu. (2012) UNDP Solution Exchange, New Delhi.

Mishra Neelkanth, Inland Fishery, Node, RRA, Net work, Pune, Maharashtra (2012), UNDP Solution Exchange, New Delhi.

Raina, J.L and Pikhan O.P. "Ecological Imbalance and agriculture productivity in the Jammu Kandi in Himalayan Environment Man and the Economic Activities (Raina Ed) (1992). Pointer Publishers, Jaipur.

ftp:/ / ftp.solution.exchange.net.in/ public / food / resources / res 14121101.docx.

Upare Maroti A, Independent Consultant Mumbai, Maharashtra (2012), UNDP Solution Exchange, New Delhi.

http:/ / www.atree.org/ matsyathavalam.

http:/ / wwwmprlp.in / downloads / TCPSU /AQ21.Aquaculture. pdf.

www.drcsc.org

□□□

4

WATER QUALITY ASSESSMENT OF SANTHEKADUR WATERBODY, SHIMOGA, KARNATAKA (INDIA) WITH RELATION TO ITS ZOOPLANKTON DIVERSITY

M. Venkateshwarlu, Shanhawaz Ahmad and K. Honneshappa

ABSTRACT

A hydro-biological study of Santhekadur pond in Shimoga district of Karnataka was carried out for a period of one year from February 2007 to January 2008, in order to access the status of plankton diversity in relation to water quality parameters. The water quality parameters and plankton diversity showed marked variations in total density, which is because of diverse hydro-biological conditions prevailing in this pond. Zooplankton diversity was represented by Cladocera, Copepoda, Rotifera and Protozoa. While, Cladocera was found to be dominant in the zooplankton diversity.

Keywords: *Zooplankton diversity, Santhekadur pond, Shimoga District, Water quality parameters.*

INTRODUCTION

Freshwater bodies comprise a vital component of the ecosystem in developing countries especially since they provide a high level of public interface. In India, there is huge number of natural and man-made water bodies used for various purpose like drinking and agriculture. However, due to various breakthroughs in the technology which lead to rapid urbanization, industrialization and modern agricultural activities, the quality of water bodies are under the stress of deterioration. Due to direct or indirect human impact, water bodies have been contaminated with variety of hazardous chemical pollutants causing an adverse impact on human health and aquatic life as well (Telliard and Rubin, 1987). Several investigators have documented and studied the hydrobiological profiles of varied lentic bodies (ponds, reservoirs, lakes) with the intention to assess the water quality (Shastri and Pendse, 2001; Azizul Islam *et al.*, 2001). The information available on the status of lentic water bodies in the Indian subcontinent shows the deterioration of water quality in general (Chandrashkhar and Jafer, 1998).

The distribution of aquatic organisms particularly plankton, has long been known to be heterogeneous among aquatic fauna. Spatial heterogeneity is a common feature in all ecosystems and is the result of many interacting physical and biological processes. The studies documented on freshwater fauna especially zooplankton, even of a particular area is extensive and complicated due to various factors to be listed (environmental, physical, geographical and chemical variations including ecological, extrinsic and intrinsic factors). Zooplankton plays a key role in the maintenance of ecological balance and its basic study is wanting and absolutely necessary. The seasonal variations of the zooplankton populations are well documented and show a bimodal oscillation with spring and autumn in the temperate lakes and reservoirs (Wetzel, 2001). This fluctuation is greatly influenced by various factors including temperature variations, which seems to be the greatest influence on the zooplankton population (Prasad, 2003).

Thus, the present study is aimed to investigate the seasonal rhythms in the physico-chemical and zooplankton diversity of the Santhekadur water body. Paradoxically enough, no work has been carried out on such an important water body and the present report is the documentation of water quality assessment with relation to its zooplankton diversity.

MATERIALS AND METHODS

Santhekadur water body is situated at Santhekaduru village and lies between latitude of 13°, 52' N and Longitude 75°, 45' E in the Shimoga city at the distance of 6 km. The total area of the tank is about 0.5 square km and has a depth of about 5 to 6 feet. It is surrounded by paddy fields and Areca plantation. It is covered with a lot of aquatic vegetation, therefore it provides protection to fish population and is the destination of many water birds also. The water body is extensively used for fishing since its inception. The major source of the water in the tank is Bhadra channel and also rainwater during rainy season. Hence, the water remains there throughout the year and therefore it is considered as perennial tank.

Field investigation was crried out for a period of one year from February 2007 to January 2008. The sampling was carried out during morning hours between 9.00 and 10.30 am. For physico-chemical analysis, water samples were collected in 1000 ml plastic bottles. The water temperature was recorded at the sampling site itself. Dissolved oxygen was fixed on the spot itself in BOD bottles. Various parameters like free CO_2, Total alkalinity, BOD, Phosphate, Nitrate, Total Hardness, Calcium, Magnesium, Total Dissolved Solids, and Chloride were estimated as per the standard methods APHA (1995).

For the qualitative and quantitative estimation of zooplankton, collections were made using a modified Haron-Trantor net with a square metallic frame of 0.0625-m_2 area. The filtering cone made up of nylon bolting silk plankton net (No. 25 mesh size 50 μ), was used for zooplankton collection. The net was hauled for a distance of 10 meters. Samples were transferred to clean 500 ml polyethylene bottles. Then, 5 ml of lugols iodine solution and 10-15 ml 4% of formaldehyde were added to it for fixation and preservation of planktonic cells. After sedimentation 100 ml of sample was subjected to centrifugation at 1500 rmp for 20 min. and used for further investigation. For the counting of zooplanktons, Sedgewick Raftar cell (Welch, 1948) was used. Identification was done by following the procedures of Edmondson (1996), Needham and Needham (1978) and APHA (1995).

RESULTS AND DISCUSSION

Physico–chemical Variables

The physico-chemical variables of Santhekadur water body is summarized in Table 1. Ambient temperature and water temperature ranges from 26.5 – 38.5 and 25.5-34.0 respectively. Mean water temperature is observed to be lower than air temperature which is attributed to less heating of the water body. This result is in conformity with Sunkad and Partil (2004). Temperature fluctuations in water temperature may be also influenced by air temperature, humidity and solar radiations. The water body was alkaline throughout the period of study and pH of the water body ranged from 7.2 to 8.5. The minimum value was noticed during monsoon period which may be due to incoming of rain water resulting in increase in turbidity, which in turn reduce photosynthetic activity of algae leading to accumulation of CO_2 and hence resulting in reduction of pH (Adibisi, 1980).

Table 1. Average values of Physico-chemical variables of Santhekadur pond.

Parameters	Pre- monsoon	Monsoon	Post- monsoon
Air Temp. (°C)	38.5	29.0	26.5
Water Temp. (°C)	34.0	28.0	25.5
pH	8.5	7.2	7.8
TDS	30.6	49.0	45.0
DO	7.8	9.5	7.2
BOD	2.0	2.41	4.35
CO_2	4.80	3.40	2.90
Chloride	30.2	14.18	11.34
Calcium	14.10	12.28	6.69
Magnesium	5.92	0.96	3.33
T. Hardness	59.0	34.6	30.5
T. Alkalinity	28.5	30.0	31.2
Phosphate	2.90	5.15	2.0
Sulphate	15.0	9.6	14.29

Note : All the parameters are in mg/L except pH and temperature (°C).

The increased pH value during summer or pre-monsoon was due to increased concentration of bicarbonate alkalinity. The present results are in conformity with the findings of Kaushik and Sharma (1994). Though, free CO_2 was present throughout the year but it did not show any seasonal surge. Dissolved Oxygen (DO) indicates physical, chemical and biological activities in a water body. It is an important indicator of water quality. DO affect the solubility and availability of many nutrients and therefore productivity of aquatic ecosystems (Wetzel, 1983). In the present study, DO values were found above the prescribed ISI value of 3 mg/l, thus giving support to the concept that under natural conditions a lentic water body typically contains a relatively higher concentration of DO tending towards saturation point (Welch, 1952). The low values of BOD indicates the low levels of biodegradable materials and absence of non-biodegradable substances. The chloride varied between 11.34-30.2 mg/l, which indicates that water appears to be suitable for irrigation purposes. It showed higher values in summer season, which could be attributed to high rate of evaporation during hotter months. A decrease trend in the chloride content in the water body during winter of post-monsoon period, may be related to the absence of dilution effect of water. Biologically important nutrient, Phosphate (PO_3) varied between 20.0 – 5.15 mg/l and showed its maximum range during rainy season indicating the influx of rain water containing fertilizers from the surrounding agricultural fields. Sulphate concentration of the water body was found to be under permissible limits and variation in sulphate content in ponds might be due to variable organic input. The results are in conformity with the results of Singh (1993). Total hardness (mg/l $CaCO_3$) and total alkalinity were found to be low and ranged from 30.5 10 to 59.0 mg/l and 28.5 to 31.2 respectively. According to Singh and Roy (1995) water with a hardness value less than 60 mg/l is soft. Hence, the present study showed that Santhekadur water body is soft and can be used for irrigation purposes. The present results are also support the work of Rawat and Jakher (2002b).

Plankton Diversity

The zooplankton density of Santhekadur water body consists of Rotifers, Cladocera, Copepoda and Ostracods (Table 2). Rotifers dominated the water body in the summer season. This may be due to higher population of bacteria, organic matter and dead and decaying vegetation. Similary, Copepods and Ostracods dominated during winter season. Summer season showed absence of Ostracods where as Copepod density was low during this period. On the other hand, Cladocerans were found to be maximum in number during monsoon period. The results are in agreement with those of Salaskar and Yeragi (2003). This shows the distinct seasonal fluctuations and composition of zooplankton in the Santhekadur water body. The variation of population of different zooplankton groups at different seasons of the year could be attributed to the availability of food material and preference towards the food in order to avoid competition (Singh, 2000). Thus, the density of different zooplankton groups are in the order of Cladocera (14 sp.)> Copepoda (7 sp.)> Rotifera (6 sp.) and Ostracods (6 sp.) respectively.

Table 2. Population density of zooplankton in Santhekadur water body.

A. Cladocera	Santhekadur
Alona pulchella	+
Ceriodaphnia cornuta	+
Daphnia cornuta	+
Daphnia carinata	+
Diaphanosoma excisum	+
Diaphanosoma sarsi	+
Macrothrix goeldi	+

Contd...

Table 2. Contd...

A. Cladocera	Santhekadur
Macrothrix laticornis	+
Maina Brachiata	+
Maina carinata	+
Alona pulchella	+
Ceriodaphnia cornuta	+
Daphnia Carinata	+
Diaphanosoma excisum	+
Diaphanosoma sarsi	+
B. Copepoda	
Heliodiaptomus sp.	+
Mesocyclops hyalinus	+
Mesocyclops leuckarti	+
Naupliar larve	+
Neodiaptomus stregilipes	+
Paracyclops fimbriatus	+
Tropocyclops prasinus	+
C. Rotifera	
Brachionus calyciflorus	+
Brachionus falcatus	+
Brachionus quadridentatus	+
Fillinia longiseta	+
Keratella tropica	+
Lepadella ovalis	+
D. Protozoa	
Arcella sp.	+
Difflugia sp.	+
Paramecium sp	+
Vorticella sp.	+

Note : (+) Present

CONCLUSIONS

It can be concluded that the variations in physico-chemical characteristics in the present study are responsible for the fluctuations in the species diversity of zooplankton of Santhekadur pond. The dominant species are reported to be the most important indicators, as they receive the full impact of the habitat and are effective tools in the environment monitoring which is required to assess the changes caused by the anthropogenic activities. Therefore, the present study based on the physico-chemical factors and zoolplankton diversity showed that these factors are within the permissible limits and hence, the water body can be considered as soft and recommended for agricultural use.

REFERENCES

Adibisi, A.A. (1980). The Physico-chemical and Hydrology of Tropical Seasonal Upper Ogun River. Hydrobiologia, 79: 157-65.

APHA, (1995). Standard Methods for the Examination of Water and Wastewater. 19 th Ed. American Public Health Association, New York, p. 1143.

Azizul, Islam M., Choudry, A.N. and Zaman, M. (2001). Limonology of Fish Ponds in Rajshahi, Bangladesh. *Ecol. Envir.* Conserve. 7: 1-7.

Chandrasekhar, S.V and Muhammed J.P. (1998). Limonological studies of a temple pond in Kerala. *Environ. Ecol.* 16: 463-367.

Edmondson, W.T. (1996). Freshwater Biology. John Wiley and Sons, Inc. New York.

Needham, J.G. and P.R. Needham. (1978). A Guide to the Study of Freshwater Biology (Holden Day Inc. Publ. San Francisco). 5th edition.

Shastri, Y and Pendse, D.C. (2001). Hydrobiological study of Dehikhuta reservoir. J. Environ. Biol. 22: 67-70.

Welch. P.S. (1948). Limnological methods. Mc Graw Hill, New York, USA.

Wetzel, R.G. (1983). Limonology. 2nd edition. Saunders Coll. Publ. p. 767.

Telliard, W.A. and M.B. Rubin. (1987). Control of pollutions in waster water. J. Chromatogr. Sci: 25, 332-327.

Prasad, N.V. (2003). Diversity and richness of zooplankton in Coringa Mangroove Ecosystem, Decadal changes. J. Aqua. Biol., Vol. 18(2):41-46.

Singh, J.P. and Roy, S.P. (1995). Limnobiotic investigation of Karvar Lake, Begusarai, Bihar. Env. Eco. 13:330-335

Kaushik, S and N. Sharma, (1994). Physico-chemical characteristics and zooplankton population of a perennial tank, matysya sarowar, Gwalior. Environ & Ecology, 12(2): 429-434.

Singh, D.N. (2000). Seasonal variation of zooplankton in tropical lake. Geobios. 27 (2-3), 97-100.

Rawat, M and Jakher, G.R. (2002b). of seasonal temperation variation on the level of dissolved oxygen, free carbon-doixide and pH in Takhat Sagar Lake, Jodhpur (Rajasthan). J. Current Sci. 2(2): 169-172.

Salaskar, P.B. and Yeragi, S. G. (2003). Seasonal fluations of plankton population correlated with physic-chemical in Powai Lake, Mumbai, Maharashta. J. Aqua. Biol. 18(1): 19-22.

□□□

5

LIMNOLOGY OF EXPERIMENTAL MACROPHYTE FLOATING ISLANDS IN A HIGH ALTITUDE RESERVOIR (VALLE DE BRAVO, MEXICO)

Pedro Ramirez-Garcia, S. Nandini *, S.S.S. Sarma,
Martha L. Gaytan-Herrera and Victor M. Almeida

ABSTACT

Valle de Bravo is a tropical high altitude (1780 m.s.l.) drinking water reservoir located in the State of Mexico (Mexico). Limnological studies are needed for this reservoir because of its importance as a source of drinking water, for aquaculture and for recreational activities. In this work we quantified the selected limnological variables (physico-chemical and zooplankton) from experimental macrophyte floating islands installed at the high altitude tropical reservoir (Valle de Bravo) in Mexico. Limnological data was collected from February to November at the Curtain (dam) sit. Water temperature varied by 6°C (from 19 to 25°C). Conductivity ranged from 160 to 187 µS/cm, depending on the site and the season. Water was generally in the alkaline range, reaching up to 9.2 during June. The dissolved oxygen levels were fairly well above normal range (7-12 mg/L) except during November at the Tizates river site. This trend was also reflected in the oxygen saturation levels. Orthophosphate levels ranged from traces to about 140 µg/L, the highest values were observed mainly during June. Total soluble phosphates were generally very low (<1 µg/L) during August and September. Similarly, total phosphates were also low (2-25 µg/L) during the same period while for the rest of the sampling periods the values on an average were three times higher. Total soluble nitrogen values were higher (average range 400-900 µg/L) throughout the sampling period, while the nitrates and nitrites were in the moderate range. Soluble ammonia levels were highest (average of about 200µg/L) during November, while the soluble organic nitrogen varied from 250 to 1600 µg/L. Total Chlorophy a levels varied from 4 to 34 µg/L. The zooplankton community was dominated by rotifers, while both, cladocerans and copepods were less diverse. The density of most zooplankton species was lower than 50 ind./L. On an average, highest mean zooplankton densities were observed at Amanalco (ca. 295 individuals /L) while the lowest was recorded at the dam site (Curtain). The zooplankton abundances were similar at the two other stations (about 80 ind./L).

Keywords: *Valle de Bravo, Mexico, Reservoir, Limnology.*

INTRODUCTION

Limnological studies in Mexico are gaining momentum with the publication of special volumes and reviews (Alcocer & Sarma, 2002; Alcocer & Brooks, 2010). Most works concentrate on large lakes and reservoirs because of their importances in drinking water, recreational and agricultural uses (De la Lanza & Garcia, 2002). These studies have given some valuable information on the range of variations in the physico-chemical variables and diversity and density of plankton. Since some of the drinking water reservoirs are also used for fish culture, information on zooplankton abundance and dynamic can be of considerable help to aquacultural operations where rotifers and cladocerans are used as diet for various species of larval fishes (Dominguez-Dominguez et al., 2002; Morales- Ventura et al., 2004).

Investigations on floating islands are of considerable interest for limnologists for various reasons. They serve as miniature wetland systems where nutrients are concentrated and thus support agricultural crops. A few countries in the world including India (Trivedy et al., 1978) and Mexico (Crossley, 2004) had used floating islands for agricultural practices. The limnological importance of floating island has been reviewed in detail by Gopal (2010). Artificial floating islands are installed in different waterbodies to monitor limnological variable around the globe (Van Duzer, 2004). However in Mexico such studies are rare.

The aim of the present work was to quantify the selected limnological variables (physico-chemical and zooplankton) from experimental marcrophyte floating islands installed at the high altitude tropical reservoir (Valle de bravo) in Mexico.

Study Area

Valle de Bravo is a tropical high altitude reservoir located west of Toluca City, in the State of Mexico, at 19°11'50" N and 100°09'913" W, and at an altitude of 1780 m.s. l (Fig. 1). It is part of the Cutzamala System located in the high basin of the Balsas River that belongs to the Hydrologic Region 18 (RH-18). The reservoir is classified as warm monomictic, stratified for nine months with anoxic hypolimnion (March to October) and, complete mixing in December (Merino-Ibarra et al., 2008). It has a suface area of about 19 km² and represents a 3.5% of the drainage basin. The mean depth (Z) is about has a sub-humid temperate climate with a rainy season during the summer (June to October), especially in July-September.

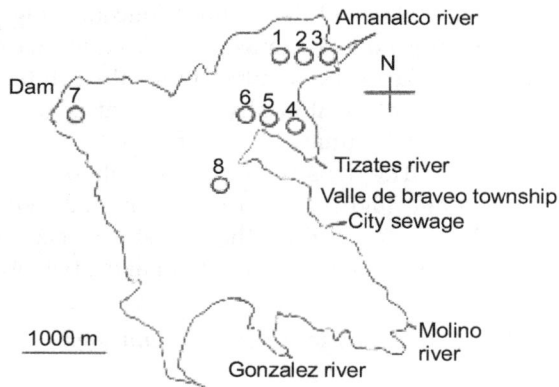

FIG. 1. Map of valle de Bravo showing sampling points. Stations 1, 2 and 3: Amanalco I, II and III; Stations 4, 5 and 6: Tizates I, II and III; Station 7: Curtain, Station 8: Centre of the reservoir.

FIG. 2. Macrophyte floating islands installed at the Valle de Bravo reservoir; Top: MFI facing Tizates River; Bottom; MFI facing Amanalco River.

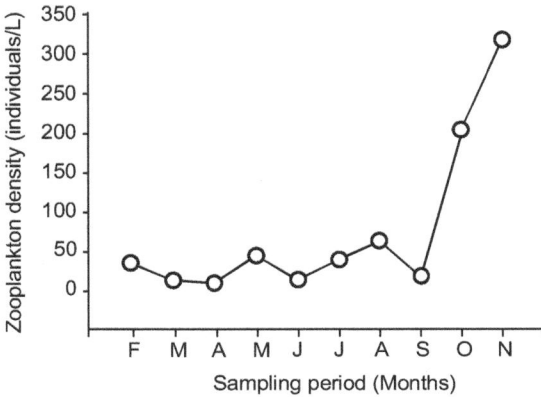

FIG. 3. Zooplankton density (individuals/L) from Macrophyte Floating Islands installed at Valle de Bravo reservoir. Shown were the mean data from different sampling stations during February to November 2010.

MATERIALS AND METHODS

In order to monitor changes in water quality we sampled two areas where we installed Macrophyted Floating Islands (MFIs) in the reservoir Valle de Bravo. From February to November 2010 we monitored the area of influence from Tizates River and Amanalco River. During dry months, for some sampling point it was not possible to collect the data and these were excluded from data analysis.

Description and location sampling stations:

> Tizates I, the nearest to the Tizates river mouth. Coordinates 19° 12,248' N and 100° 08 442' W.

> Tizates II, at a distance of 50-70 m in a straight line from the previous poin.

> Coordinates 19° 12,262' N and 100° 08 467' W.

> Tizates III, at a distance of 100-150 m straight Tizates station I. Coordinates 19° 12,288' N and 100° 08 498' W.

> Amanalco I, the nearest to the Amanalco river mouth. Coordinates 19° 13, 145' N and 100° 08 27' W.

> Amanalco II, at a distance of 50-70 m in a straight line from the previous point. Coordinates 19° 13,122' N and 100° 08 171' W.

> Amanalco III, at a distance of 100-150 m straight from Amanalco station I. Coordinates 19° 13,105' N and 100° 08 216' W.

> We selected two stations away from the shore and deep that served as the benchmark for areas without direct influence of river or effluent flows.

> Curtain, about 25 m of the wall of the dam. Coordinates 19° 12,493' N, 100° 10,794' W.

> Center of the reservoir. Around the central area of the reservoir. Coordinates 19° 11,496' N, 100° 09,191' W.

The location of the eight stations in the reservoir is presented in Fig. 1. The sampling frequency was monthly, carried out during early morning. Using a manual boat, we collected surface water. The variables considered *in situ* were: Geo-location of the sampling stations (with GPS Garmin E-Trex Legend, referred to WGS 84), temperature (°C), conditions of the day, depth at each station (Zm, m), secchi Disk Transparency (ZDS, meters, specific electrical conductivity at 25°C (K25, mS cm/L), pH, dissolved oxygen (Do, mg/L) and precent Saturation of Dissolved Oxygen (SOD,%) with calibrated YSI multiparameter probe 85.

Using the standard techniques (APHA, 1998) we estimated the following parameters in the laboratory: total phosphorus, orthophosphate, total N, nirate-N, nitrite-N, ammonia-N, organic-N, total solids, suspended solids, settled solids, biochemical oxygen demand (BOD_5), chemical oxygen demand (COD), total coliform count, fecal coliform, total bacteria (DAPI), chlorophyll a and phytoplankton (settling Utermohl chamber).

We used different forms of preservation of the samples for subsequent analysis. For measuring nitrogen, the samples were preserved using sulfuric acid and for the phytoplankton analysis, the samples were fixed in Lugol 1%, and the rest was refrigerated at 4°C until further analysis. The phytoplankton biomass was also determined using chlorophyll a extraction method. A fluorometer (10-AU Turner Designs/Perkin Elmer Lamba 25 was used to quantify the Chlorophyll a concentration.

Zooplankton samples were obtained by filtering 50 L of water from Macrophyte Floating Islands through a conical plankton mesh of 63 µm pore size. From each sampling size we filtered 90 L of water in duplicates. Zooplankton samples were concentrated to a volume of 200 mL and immediately fixed with formaldehyde at a final concentration of 4%. Some live samples were also collected for qualitative analysis.

Both phytoplankton and zooplankton samples were analyzed qualitatively using standard literature under stereo and compound microscopes. For quantitative analysis we used Sedgwick rafter cell and an inverted microscope.

RESULTS & DISCUSSION

Data on the selected physico-chemical variable of water collected during the study period are presented in Table 1. The highest Secchi transparency (nearly 5 m) was observed during November at the Curtain site. During the study period the water temperature varied little (from 19°C to 25°C). Conductivity ranged from 160 to 187 µ S cm⁻¹, depending on the site and the season. The water from the Macrophyte Floating Islands was generally on the alkaline range range and reaching up to 9.2 during June. The dissolved oxygen levels were fairly well above 6 (7-12 mg/L) except at during November at the Tizates river site. This trend was also reflected in the oxygen saturation levels.

Table 1. Data on the selected physico-chemical variables of Macrophyte Floating Islands installed at Valle de Bravo reservoir.

Secchi Transparency (m)										
	Feb	Mar	Apr	May	Jun	Jul	Aug	Sep	Oct	Nov
Tizates I	1.2	0.9	0.7	1.0	0.8	0.9	0.9	1.1	0.9	1.6
Tizates II	1.4	1.1	1.0	1.0	0.9	1.0	1.0	1.2	1.2	2.3
Tizates III	1.5	1.2	0.9	1.0	1.2	0.9	0.9	1.1	1.2	2.4
Curtain	1.4	1.2	1.1	0.9	1.3	1.5	1.0	1.5	1.4	4.9
Centre	1.5	1.0	1.0	1.1	1.2	1.4	1.1	1.3	1.4	4.5
Amanalco I	—	—	—	—	0.0	0.4	1.1	1.2	1.0	1.8
Amanalco II	—	—	—	—	0.4	0.6	1.1	1.6	1.1	2.5
Amanalco III	—	—	—	—	0.9	1.0	0.9	1.7	1.4	2.1
Temperature (°C)										
Tizates I	19.4	21.2	23.2	24.9	25.3	24.9	24.0	22.9	23.4	21.0
Tizates II	19.5	20.8	22.5	24.6	24.7	24.3	23.9	22.7	22.7	21.1
Tizates III	19.5	20.6	22.3	24.4	24.3	24.0	23.8	22.7	23.0	20.8
Curtain	18.5	19.6	21.1	23.5	23.4	23.0	23.0	22.3	21.6	19.7
Centre	18.7	19.7	21.1	23.8	23.5	23.4	23.4	22.6	22.1	19.7
Amanalco I	—	—	—	—	20.7	17.8	23.8	22.8	22.5	20.9
Amanalco II	—	—	—	—	24.9	23.4	23.7	22.9	22.5	21.0
Amanalco III	—	—	—	—	24.9	24.0	23.7	23.1	22.8	20.5

Contd...

Table 1. Contd...

Conductivity at 25°C (K25, µS cm⁻¹)										
Tizates I	180.0	178.0	184.0	185.0	181.0	175.0	170.0	166.0	167.0	177.0
Tizates II	176.0	174.0	180.0	182.0	177.0	172.0	168.0	166.0	164.0	174.0
Tizates III	173.0	173.0	179.0	181.0	177.0	172.0	167.0	163.0	163.0	17.3.0
Curtain	171.0	173.0	178.0	177.0	175.0	170.0	165.0	161.0	161.0	167.0
Centre	170.0	172.0	177.0	177.0	174.0	169.0	164.0	160.0	160.0	166.0
Amanalco I	—	—	—	—	187.0	183.0	165.0	164.0	164.0	168.0
Amanalco II	—	—	—	—	176.0	172.0	166.0	163.0	163.0	169.0
Amanalco III	—	—	—	—	177.0	172.0	166.0	163.0	162.0	169.0
pH										
Tizates I	8.4	8.7	8.5	8.7	8.9	7.8	8.7	8.6	8.7	7.2
Tizates II	8.2	8.9	8.6	8.8	9.1	8.9	8.3	8.5	8.4	7.2
Tizates III	8.6	8.9	8.7	8.9	9.1	8.8	8.2	8.6	8.5	7.2
Curtain	8.7	9.0	8.9	9.0	9.1	8.9	8.8	8.6	8.4	7.1
Centre	8.7	9.1	8.9	9.0	9.1	8.8	8.8	8.7	8.4	7.0
Amanalco I	—	—	—	—	7.4	7.0	8.9	8.7	8.5	7.5
Amanalco II	—	—	—	—	9.2	8.8	9.0	8.7	8.5	7.3
Amanalco III	—	—	—	—	9.1	8.9	8.9	8.6	8.5	7.3
Dissolved oxygen (mg/L)										
Tizates I	7.9	6.8	6.6	7.3	6.8	10.8	8.0	6.5	7.8	4.2
Tizates II	8.6	7.7	7.0	7.7	8.4	9.6	8.6	6.0	7.9	3.9
Tizates III	9.7	8.0	7.7	8.1	8.4	9.4	8.6	6.6	8.3	3.6
Curtain	9.9	7.8	8.6	8.8	8.6	9.1	8.7	6.9	7.3	3.5
Centre	9.8	8.3	8.5	8.8	8.3	9.6	8.7	7.7	7.8	3.5
Amanalco I	—	—	—	—	7.9	9.7	9.0	7.4	9.1	6.8
Amanalco II	—	—	—	—	10.3	12.1	9.8	7.2	8.9	5.5
Amanalco III	—	—	—	—	9.2	10.2	9.0	7.0	8.1	6.0
Dissolved oxygen saturation (%)										
Tizates I	85.0	77.0	74.0	85.0	85.0	133.0	94.0	75.0	91.0	47.6
Tizates II	93.0	86.0	80.0	93.0	100.0	115.0	102.0	69.0	92.0	44.2
Tizates III	106.0	89.0	89.0	97.0	100.0	112.0	102.0	76.0	96.0	40.4
Curtain	106.0	86.0	97.0	104.0	100.0	105.0	101.0	79.0	83.0	37.8
Centre	105.0	90.0	96.0	104.0	98.0	112.0	101.0	89.0	89.0	38.5
Amanalco I	—	—	—	—	87.0	102.0	107.0	86.0	105.0	74.8
Amanalco II	—	—	—	—	125.0	143.0	116.0	84.0	103.0	61.6
Amanalco III	—	—	—	—	111.0	123.0	106.0	82.0	95.0	66.6

Information on the nutrient and Chlorophyll a concentration of water collected from different sites of the macrophyte floating islands are shown in the Table 2.

Table 2. Data on the selected nutrients and Chlorophyll *a* concentration of Macrophyte Floating Islands installed at Valle de Bravo reservoir.

Ortho phosphate (µg/L)

	Feb	Mar	Apr	May	Jun	Jul	Aug	Sep	Oct	Nov
Tizates I	21.1	41.7	7.6	0.6	84.2	6.9	0.6	0.6	0.6	52.0
Tizates II	16.6	40.8	0.6	14.8	37.5	8.5	0.6	0.6	0.6	8.5
Tizates III	18.4	22.9	0.6	0.6	23.0	10.1	0.6	0.6	3.7	15.0
Curtain	14.8	19.3	0.6	14.8	32.7	8.5	0.6	0.6	0.6	0.6
Centre	17.5	18.4	0.6	14.8	37.5	5.3	0.6	0.6	0.6	0.6
Amanalco I	—	—	—	—	139.0	60.1	0.6	0.6	2.1	13.4
Amanalco II	—	—	—	—	35.9	13.4	0.6	0.6	0.6	15.0
Amanalco III	—	—	—	—	47.2	8.5	0.6	0.6	5.3	19.8

Total soluble phosphate (µg/L)

	Feb	Mar	Apr	May	Jun	Jul	Aug	Sep	Oct	Nov
Tizates I	37.3	60.3	22.4	0.6	93.9	37.5	0.6	0.6	69.7	81.0
Tizates II	32.3	59.3	0.6	30.3	40.7	34.3	0.6	0.6	48.8	31.1
Tizates III	34.3	39.3	0.6	0.6	37.5	42.4	0.6	0.6	29.5	34.3
Curtain	30.3	35.3	0.6	30.3	56.9	18.2	0.6	3.7	19.8	18.2
Centre	33.3	34.3	3.4	30.3	29.5	11.8	0.6	5.3	39.1	6.9
Amanalco I	—	—	—	—	168.0	77.8	0.6	0.6	40.7	18.2
Amanalco II	—	—	—	—	110.0	27.9	0.6	0.6	24.6	34.3
Amanalco III	—	—	—	—	73.2	31.1	0.6	0.6	23.0	25.3

Total phosphorus (µg/L)

	Feb	Mar	Apr	May	Jun	Jul	Aug	Sep	Oct	Nov
Tizates I	118.1	120.1	37.4	47.4	139.0	44.0	13.4	11.8	87.5	119.7
Tizates II	195.9	87.2	29.4	46.2	95.5	35.9	5.3	8.5	53.6	52.0
Tizates III	76.2	45.3	19.4	69.2	58.5	113.2	10.1	10.1	116.4	53.6
Curtain	39.3	78.2	1.4	34.3	77.8	26.3	15.0	18.2	68.1	29.5
Centre	54.3	42.3	31.4	31.3	42.4	16.6	29.5	24.6	44.0	24.6
Amanalco I	—	—	—	—	206.6	153.5	8.5	6.9	58.5	52.0
Amanalco II	—	—	—	—	166.4	103.6	6.9	5.3	37.5	150.3
Amanalco III	—	—	—	—	84.2	64.9	3.7	2.1	168.0	44.0

Total soluble nitrogen (µg/L)

	Feb	Mar	Apr	May	Jun	Jul	Aug	Sep	Oct	Nov
Tizates I	529	1102	1868	1180	614	436	33	297	478	626
Tizates II	507	751	1439	560	78	451	359	404	317	560
Tizates III	343	926	491	429	220	683	1144	808	304	546
Curtain	443	601	540	541	315	58	312	381	286	477

Contd...

Table 2. Contd...

Centre	332	690	579	910	410	386	470	274	307	470
Amanalco I	—	—	—	—	1366	1344	907	539	534	427
Amanalco II	—	—	—	—	1199	711	485	719	353	398
Amanalco III	—	—	—	—	405	619	1659	1285	443	413
Soluble nitrates (µg/L)										
Tizates I	0.5	0.5	0.5	0.5	26.7	12.0	0.5	0.5	2.7	8.1
Tizates II	0.5	0.5	0.5	0.5	18.2	11.2	0.5	0.5	0.5	3.0
Tizates III	0.5	0.5	0.5	9.0	43.6	36.1	0.5	0.5	0.5	1.7
Curtain	20.0	0.5	0.5	10.0	0.5	0.5	0.5	8.0	0.5	0.5
Centre	30.0	0.5	0.5	8.5	0.5	0.5	0.5	0.5	0.5	0.5
Amanalco I	—	—	—	—	958.5	553.6	215.0	0.5	0.5	13.0
Amanalco II	—	—	—	—	778.4	133.7	60.0	0.5	0.5	12.1
Amanalco III	—	—	—	—	15.4	10.5	0.5	0.5	0.5	21.2
Soluble nitrites (µg/L)										
Tizates I	4.0	14.0	—	5.0	7.0	5.0	6.0	2.0	6.0	19.0
Tizates II	5.0	15.0	—	4.0	2.0	2.0	2.0	3.0	4.0	19.0
Tizates III	2.0	11.0	—	3.0	2.0	2.0	2.0	2.0	3.0	18.0
Curtain	2.0	12.0	—	3.0	3.0	2.0	2.0	9.0	4.0	16.0
Centre	2.9	12.0		2.0	3.0	3.0	8.0	2.0	3.0	19.0
Amanalco I	—	—	—	—	18.0	143.0	150.0	2.0	5.0	17.0
Amanalco II	—	—	—	—	20.0	2.0	2.0	2.0	6.0	18.0
Amanalco III	—	—	—	—	2.0	2.0	2.0	2.0	1.0	18.0
Soluble ammonia (µg/L)										
Tizates I	225.0	238.0	268.0	115.0	81.0	29.1	27.0	17.0	130.0	349.0
Tizates II	122.0	66.0	189.0	166.0	60.0	50.4	29.0	11.0	63.0	288.0
Tizates III	41.0	95.0	141.0	87.0	175.0	87.6	34.0	18.0	51.0	277.0
Curtain	91.0	39.0	190.0	78.0	32.0	30.0	32.0	24.0	31.0	210.0
Centre	30.0	58.0	229.0	60.0	77.0	53.1	22.0	14.0	23.0	200.0
Amanalco I	—	—	—	—	109.0	147.4	42.0	29.0	138.0	147.0
Amanalco II	—	—	—	—	121.0	77.1	35.0	39.0	96.0	118.0
Amanalco III	—	—	—	—	58.0	46.0	49.0	45.0	101.0	124.0
Soluble organic nitrogen (µg/L)										
Tizates I	300	850	1600	1060	500	390	240	280	340	250
Tizates II	380	670	1250	390	250	390	330	390	250	250
Tizates III	300	820	350	330	250	560	1110	790	250	250
Curtain	330	550	350	450	280	28	280	340	250	250

Contd...

Table 2. Contd...

Centre	270	620	350	840	330	330	440	260	280	250
Amanalco I	—	—	—	—	280	500	500	510	390	250
Amanalco II	—	—	—	—	280	500	390	680	250	250
Amanalco III	—	—	—	—	330	560	1610	1240	340	250

Chlorophyll *a* concentration (μg/L)

Total Chlorophyll *a*

Tizates I	18.5	33.9	20.5	11.5	6.8	13.3	9.0	6.5	7.3	18.5
Tizates II	29.9	25.2	27.3	9.6	30.5	10.9	11.6	8.8	8.1	29.9
Tizates III	27.5	33.1	17.6	9.7	6.0	13.2	9.1	8.7	7.0	27.5
Curtain	27.0	25.3	18.5	17.3	5.2	8.4	9.6	5.5	9.1	27.0
Centre	21.3	24.3	15.5	11.7	4.2	12.5	9.9	5.1	5.5	21.3
Amanalco I	—	—	—	—	—	—	—	—	—	—
Amanalco II	—	—	—	—	—	—	—	—	—	—
Amanalco III	—	—	—	—	13.3	16.3	12.8	3.9	11.8	—

Chlorophyll *a* from <20 μm fraction

Tizates I	6.7	25.8	12.2	6.4	7.4	3.4	4.5	3.7	4.5	18.5
Tizates II	11.6	22.6	17.9	4.1	1.9	3.5	5.2	0.5	6.9	29.9
Tizates III	9.5	20.4	12.2	4.2	2.6	3.6	6.8	4.1	4.8	27.5
Curtain	17.0	21.3	16.4	5.2	2.8	2.5	2.2	4.1	1.4	27.0
Centre	17.4	17.6	10.7	3.5	2.0	2.2	2.4	0.4	2.5	21.3
Amanalco I	—	—	—	—	—	—	—	—	—	—
Amanalco II	—	—	—	—	—	—	—	—	—	—
Amanalco III	—	—	—	—	2.9	2.8	5.3	1.2	5.5	—

Orthophosphate levels ranged from traces to about 140 μg/L, the highest values were observed mainly during June. Total soluble phosphate were generally very low (<1 μg/L) during August and September. Similarly, total phosphates were also low (2-25 μg/L) during the same period while for the rest of the sampling periods the values on an average were three times higher. Total soluble nitrogen values were higher (average range 400-900 μg/L) throughout the sampling period, while the nitrates and nitrites were on the moderate range. Soluble ammonia levels were highest (average of about 1600 μg/L) during November, while the soluble organic nitrogen varied from 250 to 1600 μg/L. Total Chlorophyll a levels varied from 4 to 34 μg/L. However, Chlorophyll a from the phytoplankton fraction of <20 μm contributed to more than 60% of the total Chlorophyll a.

Data on the total solids, suspended solids, dissolved solids and organic matter from the suspended solids are provided in Table 3. The total solids varied from 85 to >200 mg/L while the suspended solids were much lower (1 to 30 mg/L). Total dissolved solids were also high (70 to 175 mg/L). The organic matter from the suspended solids ranged from 1 to 14 mg/L.

Table 3. Data on the solids (dissolved and suspended) of Macrophyte Floating Islands installed at Valle de Bravo reservoir.

Total solids (mg/L)

	Feb	Mar	Apr	May	Jun	Jul	Aug	Sep	Oct	Nov
Tizates I	155.0	107.5	175.0	112.5	110.0	165.0	125.0	105.0	110.0	100.0
Tizates II	177.5	97.5	92.5	112.5	115.0	145.0	80.0	85.0	95.0	100.0
Tizates III	160.0	102.5	147.5	100.0	107.5	155.0	130.0	105.0	95.0	90.0
Curtain	180.0	92.5	127.5	100.0	110.0	150.0	115.0	95.0	110.0	100.0
Centre	180.0	107.5	132.5	92.5	85.0	105.0	100.0	110.0	100.0	100.0
Amanalco I	—	—	—	—	135.0	205.0	1 35.0	125.0	90.0	85.0
Amanalco II	—	—	—	—	100.0	160.0	115.0	115.0	110.0	85.0
Amanalco II	—	—	—	—	85.0	175.0	115.0	120.0	95.0	90.0

Total suspended solids (mg/L)

	Feb	Mar	Apr	May	Jun	Jul	Aug	Sep	Oct	Nov
Tizates I	6.0	7.5	10.2	6.7	14.0	7.0	6.3	8.0	9.3	1.3
Tizates II	7.8	6.8	9.8	8.0	12.7	4.3	8.0	5.0	5.3	0.8
Tizates III	8.3	6.8	10.7	10.0	15.7	4.7	7.3	6.7	6.3	2.5
Curtain	7.0	6.8	12.8	5.3	8.7	5.0	8.3	5.7	3.3	1.8
Centre	7.5	10.5	7.4	5.0	8.3	6.7	8.0	5.0	4.7	2.5
Amanalco I	—	—	—	—	21.7	30.7	6.7	6.3	8.0	1.0
Amanalco II	—	—	—	—	15.7	14.0	8.7	5.3	7.3	1.0
Amanalco	—	—	—	—	15.7	7.7	8.3	4.7	7.3	1.7

Organic matter from suspended solids (mg/L)

	Feb	Mar	Apr	May	Jun	Jul	Aug	Sep	Oct	Nov
Tizates I	2.8	3.3	6.5	3.8	7.7	3.3	2.7	4.7	5.7	1.3
Tizates II	2.3	2.0	7.6	1.7	6.7	0.7	5.0	3.7	2.3	0.7
Tizates III	3.5	1.8	8.3	5.8	14.0	2.7	4.3	5.0	3.0	2.5
Curtain	2.0	2.3	8.9	3.8	7.8	2.3	4.0	4.3	2.7	1.8
Centre	2.5	2.5	5.7	4.2	8.0	4.3	6.0	3.7	2.0	2.3
Amanalco I	—	—	—	—	5.7	7.3	2.3	4.0	3.0	1.0
Amanalco II	—	—	—	—	8.3	4.3	4.7	5.0	3.7	1.0
Amanalco III	—	—	—	—	7.3	1.3	6.3	3.3	3.3	1.7

Total dissolved solids (mg/L)

	Feb	Mar	Apr	May	Jun	Jul	Aug	Sep	Oct	Nov
Tizates I	149.0	100.0	164.8	105.8	96.0	158.0	118.7	97.0	100.7	98.7
Tizates II	169.8	90.8	82.7	107.7	102.3	140.7	72.0	80.0	89.7	99.2
Tizates III	151.8	95.8	136.9	90.0	91.8	150.3	122.7	98.3	88.7	87.5
Curtain	173.0	85.8	114.7	94.7	101.3	145.0	106.7	90.0	106.7	98.2
Centre	172.5	97.0	125.1	87.5	77.7	98.4	92.0	105.0	95.3	97.5 ·
Amanalco I	—	—	—	—	113.3	174.3	128.3	118.7	82.0	84.0
Amanalco II	—	—	—	—	84.3	146.0	101.3	109.7	102.7	84.0
Amanalco III	—	—	—	—	69.3	167.3	107.0	115.3	87.7	88.3

Total bacterial density, total coliform and fecal coliform bacterial densities are given in Table 4.

Table 4. Data on the total bacteria and fecal bacterial densities from Macrophyte Floating Islands installed at Valle de Bravo reservoir.

Total bacteria (X10^6/ml)

	Feb	Mar	Apr	May	Jun	Jul	Aug	Sep	Oct	Nov
Tizates I	14	8	14	7	19	24	19	6	4	7
Tizates II	13	4	9	6	11	17	14	11	3	8
Tizates III	10	9	10	7	13	12	14	10	1	6
Curtain	6	6	12	9	5	14	19	9	1	5
Centre	6	6	6	8	15	17	17	8	4	5
Amanalco I	—	—	—	—	8	15	18	4	1	9
Amanalco II	—	—	—	—	13	20	1	1	4	9
Amanalco III	—	—	—	—	15	15	9	5	4	7

Total Coliform bacteria (no./ml)

	Feb	Mar	Apr	May	Jun	Jul	Aug	Sep	Oct	Nov
Tizates I	3500	2400	170	10	10.0	130	11	5.4	540	5.4
Tizates II	170	160	170	6.0	1.0	49	24	7.0	240	9.2
Tizates III	920	54	50	0.8	2.0	35	24	8.0	13	3.5
Curtain	35	3.5	2.3	0.1	0.1	2.8	2.2	1.1	3.5	0.1
Centre	5.4	0.2	1.3	0.2	0.1	1.7	0.5	0.8	0.8	0.1
Amanalco I	—	—	—	—	9.0	920	2.8	3.3	3.3	0.4
Amanalco II	—	—	—	—	0.4	54	9.2	2.7	1.3	0.1
Amanalco III	—	—	—	—	0.5	7.9	16	2.1	1.7	0.3

Fecal Coliform bacteria (no./ml)

	Feb	Mar	Apr	May	Jun	Jul	Aug	Sep	Oct	Nov
Tizates I	3500	2400	130	10	10	49	11	2.2	540	2.2
Tizates II	27	35	22	6.0	1.0	49	24	4.9	240	9.2
Tizates III	280	54	50	0.6	1.0	35	24	8	7.9	3.5
Curtain	16	3.5	0.2	0.1	0.1	2.8	2.2	0.8	1.7	0.1
Centre	0.5	0.2	0.2	0.1	0.1	1.4	0.5	0.2	0.5	0.1
Amanalco I	—	—	—	—	9.0	350	2.2	1.3	1.3	0.2
Amanalco II	—	—	—	—	0.4	54	9.2	0.2	0.6	0.1
Amanalco II	—	—	—	—	0.2	0.8	16	2.1	1.7	0.2

At the sampling stations of Tizates, there were higher bacterial abundances mainly during October, November and February.

Table 5. Data on the BODs (mg/L) and COD from Macrophyte Floating Islands installed at Valle de Bravo reservoir.

(BOD) Biological Oxygen Demand (mg/L)

	Feb	Mar	Apr	May	Jun	Jul	Aug	Sep	Oct	Nov
Tizates I	1.6	3.0	2.6	3.6	3.6	2.2	3.0	3.1	3.6	3.8
Tizates II	1.1	2.3	2.2	3.6	2.3	2.2	3.3	2.7	2.5	1.6
Tizates III	1.3	2.3	2.1	3.2	2.3	1.6	3.2	2.6	2.3	2.7
Curtain	0.8	2.7	1.5	2.3	1.8	1.9	2.9	2.1	1.8	0.8
Centre	1.4	2.6	1.8	2.5	1.7	1.2	2.5	1.5	1.9	1.2
Amanalco I	—	—	—	—	1.1	1.0	1.7	2.3	3.9	3.1
Amanalco II	—	—	—	—	1.5	4.4	2.8	2.6	3.2	2.4
Amanalco III	—	—	—	—	1.2	2.6	3.8	2.9	2.8	3.2

(COD) Chemical oxygen demand (mg/L)

	Feb	Mar	Apr	May	Jun	Jul	Aug	Sep	Oct	Nov
Tizates I	9.8	15.6	21.1	21.3	22.1	16.8	12.1	19.0	15.9	11.1
Tizates II	11.7	15.2	17.9	16.4	21.7	11.7	11.1	12.7	11.1	9.5
Tizates III	8.9	16.5	18.3	15.9	18.3	16.5	9.2	12.7	15.9	6.3
Curtain	9.2	12.1	17.5	17.6	25.9	9.5	6.7	7.9	11.1	4.8
Centre	7.9	12.1	16.4	14.0	14.6	12.4	12.1	9.5	12.7	7.9
Amanalco I	—	—	—	—	23.5	13.0	9.2	14.3	11.1	8.6
Amanalco II	—	—	—	—	26.0	14.3	22.2	14.3	14.3	4.1
Amanalco III	—	—	—	—	34.4	23.2	14.4	14.3	79	6.3

The zooplankton community was dominated by rotifers, while both cladocerans and copepods were less diverse. The density of most zooplankton species was lower than 50 ind./L. On an average, highest mean zooplankton densities were observed at Amanalco (ca. 295 individuals / L) while the lowest was recorded at the dam site (Curtain). The zooplankton abundances were similar at the two other stations (about 80 ind./L) (Table 6).

Table 6. Zooplanktons (Rotifera, Cladocera and Copepoda) mean abundances (ind./L1 from different sites of the Macrophyte Floating Islands installed at Valle de Bravo reservoir.

Species	Tizates	Curtain	Centre	Amanalco
Rotifera				
Anuraeopsis fissa (Gosse)	<1	0	<1	<1
Ascomorpha ovalis (Bergendal)	11	00		
Asplanchna priodonta (Gosse)	<1	<1	10	
Brachionus havanaensis (Rousselet)	00	01		
Collotheca sp.	1	<1	1	2
Euchlanis dilatata Ehrenberg	<1	0	0	0

Contd...

Table 6. Contd...

Filinia longiseta (Zacharias)	<1	0	1	0
Kellicottia bostoniensis (Rousselet)	0	0	1	0
Keratella cochlearis (Gosse)	46	9	27	133
K. americana Carlin	<1	0	1	8
Lecane bulla (Gosse)	<1	0	0	0
L. closterocera (Schmarda)	1	0	0	0
Platyias quadricornis (Ehrenberg)	1	2	1	2
Polyarthra dolichoptera Idelson	1	0	0	0
P. vulgaris Carlin	12	11	21	78
Pompholyx sulcata (Hudson)	<1	<1	1	0
Synchaeta pectinata Ehrenberg	1	2	4	0
Trichocerca capucina (Wierzejski & Zacharias)	<1	1	4	0
T pusilla (Lauterborn)	<1	0	<1	0
T. similis (Wierzejski)	<1	2	1	<1
Trichotria tetractis (Ehrenberg)	<1	0	0	0
Cladocera				
Alona sp.	<1	<1	1	0
Bosmina longirostris (Mueller)	11	33	12	40
Ceriodaphnia lacustris Birge	1	1	0	4
Chydorus sphaericus (Mueller)	1	3	2	9
Daphnia ambigua Scourfield	0	0	0	0
D. laevis Birge	<1	0	0	0
Copepoda				
Copepod nauplii	5	2	2	12
Cyclopoid males	<1	<1	<1	0
Cyclopoid females	1	<1	<1	6
Calanoid males	<1	0	0	<1
Calanoid females	<J	<J	0	0

This is one of few studies that considers detailed limnological and hydrobiological variables. Thus this study adds additional information to the existing data collected from the reservoir at different depths. The magnitude of variables recorded from the macrophyte floating islands were comparable to those taken from water column from previous studies. For example, the magnitude of variations in Secchi transparency, temperature, conductivity, pH and dissolved oxygen levels is similar to those recorded in previous studies (Ramirez-Garcia et al., 2002; Nandini *et al.*, 2007; 2008; Jimenez Contreras *et al.*, 2009).

Detailed physico-chemical variables of this reservoirs have been documented by Merino-Ibarra *et al.*, (2008). According to these authors, this reservoir has developed a tendency towards eutrophication. They have observed that chlorophyll a was low during mixing (9 µg/L), and high (up to 88 µg/L) during stratification. In the present work too the Chlorophyll a varied from 5 to 35 µg/L, depending on the Sampling period and the Station.

Valle de Bravo reservoir is characterized by high rotifer dominance. We recorded in our MFI installations as many as 21 rotifer species. This has also been observed in previous works. For example, Jimenez-Contreras et al., (2009) have reported 23 rotifer species of which the genera *Keratella, Polyarthra* and *Trichocerca* dominant constituting about 80% of the total numerical abundances. They have also observed the highest rotifer density during April (<1600 individuals/L) and the lowest was during January (<50 ind./L). In the present work the total zooplankton abundances varied from 25 to 325 ind./L. In addition, we also observed numerical dominance of *Keratella, Polyarthra* and *Trichocerca.* We did not observe high abundance of *Anuraeopsis* possibly due to the fact that this genus is a typical planktonic and our MFI were probably not an appropriate niche for it.

The near absence of *Brachionus* in this reservoir has been commented upon ealier (Ramirez-Garcia *et al.*, 2002; Nandini et al., 2007). In this work too we observed only one species of *Brachionus.* Low temperature and absence of appropriate niches are probably responsible for this (Sharma & Sharma, 2005). As in the previous works (Nandini *et al.*, 2008; Jimenez-Contreras *et al.*, 2009), the dominant rotifer species were *Keratella chochlearis, Polyarthra vulgaris* and *Trichocerca similis.* Reports on the density and diversity of Crustacean zooplankton from this reservoir indicate occurrences of *Bosmina, Daphnia, Ceriodaphnia* and a few species from Calanoidea and Cyclopoidea (Ramirez-Garcia et al., 2002). We also observed these taxa in our present work. *Bosmina longirostris* was recorded in highest abundance (40 ind./L) from the Amanalco sampling station.

We found that in the presence of the floating islands in Amanalco the nutrient levels were negligible yet the zooplankton densities were highest. Further studies are needed in this reservoir in order to confirm an improvement in water quality as a result of high densities of macrophytes in the littoral region.

ACKNOWLEDGEMENTS

The work was supported by a project PAPCA (2010-11FES Iztacala,#24) and CAN/ UNAM_No. OAVM-DT-MEX-10-440-RF-CC. SN and SSSS also thank FES Iztacala for support extended to members SNI-CONACYT and PAPIIT-IN221111. The authors thank B.L.Kaul for helpful comments.

REFERENCES

Alcocer J, Bernal-Brooks F W 2010. Limonolgy in Mexico. Hydrobiologia 644: 1-54.

Alcocer J, Sarma SSS (Eds) 2002. Advances in Mexican Limnology: Basic and Applied Aspects. Hydrobiologia 467: 1-228 pp.

APHA 1998. Standard Methods for the Examination of water and Wastewater. 20 th Ed. Washington, DC.

Crossley PL 2004. Just beyond the eye: floating gardens in Aztec Mexico. Historical Geography 42: 111-135.

De la Lanza EG, Garcia CJL 2002. Lagos y presas de Mexico. AGT Editor S.A., Mexico DF: 1, p. 680.

Dominguez-Dominguez O, Nandini S, Sarma SSSS 2002. Larval feeding behaviour of the endangered fish golden bubblebee goodied, Allotoca dugesi (Bean) (Goodeidae) offered zooplanktron: implications for conservation. Fish. Manag. Ecol. 9: 285-291.

Gopal B 2010. Conservation and Management of lakes. An Indian perspective. Ministry of Environment & Forests, Government of India. New Delhi: pp. 1-101.

Jimenez-Contreras J, Sarma SSS, Merino-lbarra M, Nandini S 2009. Seasonal changes in the rotifer (Rotifera) diversity from a tropical high altitude reservoir (Valle de Bravo, Mexico). Journal of Environmental Biology 30: 191-195.

Merino-Ibarra M, Monroy-Rios E, Vilaclara G, Castillo FS, Gallegos ME, Ramirez- Zierold J 2008. Physical and chemical limnology of a wind-swept tropical highland reservoir. Aquatic Ecology 42:335-345.

Morales-Ventura J, Nandini S, Sarma SSS 2004. Functional responses during the early larval stages of the charal fish Chirostoma riojai (Pisces: Atherinidae) fed Zooplankton (rotifers and cladocerans). Journal of Applied Ichthyology 20: 417-421

Nandini S, Merino-Ibarra M, Sarma SSS 2008. Seasonal changes in the zooplankton abundances of the reservoir Valle de Bravo (State of Mexico, Mexico). Lake and Reservoir Management 24: 321-330.

Nandini S, Sarma SSS, Ramirez-Garcia P 2007. Seasonal variations of zooplankton from a drinking water reservoir (Valle de Bravo) in Mexico. Chapter 8. In: B. L. Kaul (Ed.). Advances in fish and wildlife ecology and biology. Vol. 4. ISBN: 8170355176. Daya Publishing House, Tri Nagar, Delhi, India: 75-86.

Olvera- Viascan V, Bravo-Inclan L, Sanchez-Chavez J 1998. Aquatic ecology and Management assessment in Valle de Bravo reservoir and its watershed. Aquat Ecosyst Health manage 1:277-290.

Ramirez-Garcia P, Nandini S, Sarma SSS, Robles-Valderrama E, Cuesta I, Hurtado-Maria D 2002. Seasonal variations of zooplankton abundances in the freshwater reservoir Valle de Bravo (Mexico). Hydrobiologia 467: 99-108.

Trivedy RK, Sharma KP, Goel PK, Gopal B 1978. Some ecological observation on floating islands. Hydrobiologia 60: 187-190.

Van Duzer C 2004. Floating Island. A Global Bibliography. Cantor Press, Loss Altos Hills, California. 1-404.

❑❑❑

6

EFFECT OF DIFFERENT NUTRIENT MEDIA ON THE RELATIVE GROWTH, BROOD SIZE AND POPULATION DYNAMICS OF *MOINA MACROCOPA*

Satinder Kour

ABSTRACT

Expansion and production consistency in commercial farming of aquatic species, particularly species of marine fish, are limited by the lack of successful replacement of live food with formulated feeds. Lack of progress may be due to insurmountable objectives created by the persistence of several biases about what constitutes the proper formulated feed and condition for larval culture. Poor response to larval diets may have nothing to do with nutrient composition, but rather the result of incompatible culture conditions and unsatisfactory presentation of the diet.

Keywords: Crustacean *Moina macrocopa*, nutrient media, population dyndimircs.

INTRODUCTION

The culture of larvae of many species of fish and crustaceans is precariously depended upon the availability of live food, whether plant or animal. The newly hatched *Moina macrocopa* generally served as an excellent source of food for larvae of many species of fish and crustanceans. However, live *Moina* are obtained through hatching of cysts that are collected from the natural environment and are subject to periodic, unpredictable shortage that cannot supply the demand. As a result, prices increase, leading to overall increases in production costs. In addition, another chronic problem is that temporal or spatial differences in cyst collections are reflected in variation in the nutritional quality of hatched *Artemia* nauplii.

Formulated diets that can achieve consistent and reliable production equivalent to that of live food still do not exist and have definitely been an impediment to the progress of marine fish culture throughout the world. During the past decade significant progress has been achieved but the common use of microdiets, characterized as either microbound, microencapsulated, or microcoated (Tucker, 1998), as exclusive sources of nutrition still eludes us and remains a challenge. Research that has addressed the development and evaluation of nutritionally complete larval diets has been conducted for over three decades. Most diets, at best, have served as supplements rather than complete substitutions (Kumlu & Jones 1995). The lack of success might in part be due to imposed biases that are partly associated with trying to realize an ideal that may not be attainable. Success may only be achieved through the creation of a diet that possesses less than the ideal characteristics. The highest level of success has been predominantly realized with herbivorous forms of crustacean larvae (Bautista, Millamena & Kanazawa, 1989; Koshio *et al.*, 1989). For these forms, rapid rates of transit time through the gut combined with the highest levels of enzyme activity (Jones, Yule & Holland 1997) provide for the satisfaction of nutrient requirements is achieved through volume of contact per unit of time rather than efficient digestion. In contrast, the comparatively short gut retention time of carnivorous larvae, particularly during early stages of metamorphosis, precludes the use of poorly digestible diets because rapid digestion is critical.

The intent of this paper is to suggest that some of the previous research is based upon biases that need to be relinquished. Successful development of micro diets for larvae in the future may only be achieved through the recognition that a useful diet must be simple to produce and that creation of the ideal diet is not possible due to needs that are often in direct conflict. The goal of developing successful diets depends on a better understanding of nutritional physiology as well as the elimination of biases. My thoughts concerting some of the most important factors and the biases that have been prevalent in the past follow:

FACTORS INFLUENCING THE EFFECTIVENESS OF FORMULATED DIETS

Enzyme Activity and Digestive Capacities

At one time, the prevailing explanation for the perennial lack of success in the culture of carnivorous larvae was insufficient enzyme activity. Researchers speculated that the enzyme manufacturing capacity within the gut was far lower that what was needed and that effective digestion was only eccomplished through the assistance of exogenous enzymes that originated from the sources of lived food (Kolkovski, Tandler & Isquierdo 1997; Kolkovski *et al.*, 1993; Munilla-Moran, Stark & Barbour 1990). However, Lovett & Felder (1990) found that contribution of enzyme activity from Artemia prey was very low compared to that measured in larvae of Penaeus setiferus. Cahu & Zambonino Infante (1997) found that the lack of good growth in sea bass larvae (15-40 days old) fed a formulated diet was not due to a lack of endogenous enzyme activity. Garcia-Ortega et al. (1998) observed a minimal contribution of enzymes. Formulated diets for larvae 145 from consumed prey to the larval gut. It appears that although the complement of digestive enzymes available might be qualitatively restricted, particularly during early stages of development, a sufficient quantitative and qualitative array of enzymes do exist to achieve good growth and survival to metamorphosis.

If the complement of enzymes exist, then why do micro diets remain ineffective? The ineffectiveness may reside in a lack of understanding of the specificity of the enzymes that exist at different stages of larval development. The digestive capacities of larvae relative to the ingredients of a formulated feed may be species of stage-specific. Changes in the sequence of qualitative enzyme activity is consistent with the strictly carnivorous feeding activity of early larval stages of the freshwater prawn *Macrobrachium rosenbergii* (Kamarudin *et al.*, 1994) and the ontogenetic change in enzyme activity has also been described for the larvae of the white shrimp *Penaeus setiferus* (Lovett & Felder, 1990). Digestive capacity corresponds to the anatomical development of the digestive system which in turn is related to changes in habitat and diet during metamorphosis (Lovett & Felder 1989). The absence of digestive capacity can be attributed to either physical or chemical characteristics that are not compatible with the enzymatic capacity of very young fish larvae. Digestive capacity may also be based upon the quality with culture of larvae of *Macrobrachium rosenbergii* using formulated diets that were produced in a variety of ways and contained egg albumin as the primary protein source were unsuccessful. (Ohs *et al.*, 1998). These diets were readily consumed and the guts were full; however, growth and survival were significantly lower than that achieved with live food. This diet was a modification of a similar egg albumin based diet that proved successful for the axenic culture of successive generations of the microcrustancean Moina macrocopa (D'Abramo, 1979). Recent efforts to feed larvae of *M. rosenbergii* with a diet wherein the primary protein source is egg yolk have proved successful (Kovalenko et al., 2002). This minor change suggests that the egg albumin protein, despite possessing an excellent amino acids profile, is not efficiently utilized. Although a sufficient quantity of enzymes may be present, they are highly specific and only effective with certain types of protein sources. Therefore, attempts to mimic the ingredient composition of larval diets with that of successful diets for juvenile culture is an approach that may not be possible. One must be aware that the large amount of published literature concerning the nutrition of juveniles may not be applicable to larval forms. Therefore, failure towards the goal of including what would be considered to be an excellent dietary protein source relative to the amino acid composition of the larvae may be due to its indigestible properties for larvae.

Physical Form of the Diet

The success of any micro diet for larval culture is critically dependent upon physical form. Successful consumption of the diet may be dependent upon the moisture content of the diet, which ultimately affects the physical form of the diet. Ideally, a diet that has low moisture content would be most desirable because it offers advantages relative to storage in a frozen Louis R. D'Abramo 146 state for an extended period of time. In contrast, high moisture diets naturally limit the duration of storage and shelf life. In addition, the rate of leaching of water soluble nutrients would presumably be amplified in formulated larval diets containing a high moisture content a dry diet has many appealing benefits but this bias based upon the overriding desire for convenience must not prevail. For dry diets, the method of drying may also influence the performance of a formulated diet. Teshima & Kanazawa (1983) found differences in growth and survival of larvae of *Penaeus japonicas* when fed carrageen an microbound diets dried under different conditions. High moisture diets may also offer a level of palatability and eventual consumption that cannot be achieved with dry diets. Shape and size are other important characteristics. The size of formulated diets designed for fish and filter feeding crustacean larvae is generally confined to particles that are small enough to be entirely consumed without the assistance of mouthparts. Larvae of sea bream have been observed to ingest inert particles that are 60 to 80% of the width of the mouth (Fernandez-Diaz, Pascual & Yufera 1994). However, this need is not always warranted.

Although diets that contain high levels of moisture may not be amenable to the production of particles less than 150 microns, restrictions of particle size imposed by high moisture formulated diets may not be applicable to some species for crustacean larvae that are raptorial feeders. These species possess mouthparts of appendages that permit the grasping and manipulating of particles that are physically modified into sizes that are consumed. Shape of the particle may also influence consumption of formulated diets. Some species may prefer irregular size particles whereas the consumption of food by other species may be limited to smooth particles. This type of work suggests that videography of the feeding of larva forms will have to become an important part of the evaluation of the consumptive "appeal" of formulated diets. The diet may have the necessary size and nutrient profile but still remain unacceptable. Considerable effort has been devoted to establishing diets that are neutrally buoyant to afford as much access to the diet in the water column. However, achieving this ideal may be unnecessary. Although some diets may have comparatively high rates of sinking, maintenance in the water column can be achieved with appropriate aeration or design of the culture tank. The lack of buoyancy may pose no threat to successful culture of some species that have been observed to feed to readily off the bottom.

The lack of ingestion of formulated diets may in fact be the lack of visual stimuli. Some of that stimuli may reside in the movement or colour of the live prey, characteristics that are note absent or not normal in formulated diets.

Frequency of Feeding

The prevailing thought is that frequent feeding is an important component of successful larval culture. Yet, Rabe & Brown (2000) found that the rate of comsumption of live food, Artemia and Brachionus sp. by larvae of yellowtail flounder was significantly higher when live food was offered 1 X or 2 X per day rather than continuously. These results suggest formulated diets for larvae 147 that the labor associated with frequent provision of formulated diets may not be necessary for some species of larvae. A more rapid consumption of food and more efficient use of nutrients may actually be achieved through a reduction in the number of feedings per day.

A reduction in the frequency of feedings may also reduce the incidence of fouling and reduce the period of time when diets remain unconsumed and subject to leaching of water soluble nutrients. The assumption that frequent feeding is essential for larvae may not be universally applicable and can create undue restrictions on the type of diet that can be used.

For example, continuous delivery of moist diets through timed belt feeders may not be possible but such a limitation is non-existent if feeding could be limited to 2-3 X per day without any adverse effects on growth and survival.

Lipid and Fatty acid Nutrition

Dietary lipid is recognized as an energy source that would seemingly be more critical to achieving satisfactory growth of carnivorous rather than herbivorous larvae. For carnivores, it is assumed that the quantity and quality of carbohydrates produced by carnivorous larvae would be limited based upon the known composition of the natural diet.

Lipid in the form of cholesterol, phospholipids, and essential fatty acids has been found to be important for the nutritional physiology of fish and crustacean larvae (Teshima, Ishikawa & Koshio 1993; Jones, Kanazawa & Rahman 1979). Fatty acid composition of the tissue of freshwater verus marine organisms is different and these differences are important considerations in the formulation of larval diets for marine verus freshwater organisms. Recently, the important role of docosahexaenoic acid was demonstrated in feeding experiments with haddock larvae using life and formulated diets (Blair et al., 2002). Certain highly unsaturated fatty acids are found in the tissue at concentrations that are disproportionate to those levels found in the diet, suggesting that these larvae have a tremendous ability to sequester these fatty acids.

Perhaps a threshold dietary level is necessary for the larvae to attain the desired amount in the tissue for normal growth, development, and survival. The idea of producing a diet that reflects as much as possible the relative amounts for macronutrients found in the natural diet may be unattainable because establishing the physical integrity of a diet often requires the addition of carbohydrates.

Gut Retention Time

The rate of passage of food through the gut of a larval fish or crustacean will definitely influence the relative effectiveness of a particular diet. Motility through the gut decreases during ontogeny and average time of food retention accordingly increase as observed in larval lobsters (Kurmarly, Jones & Yule 1990). This information should be an important source of guidance in developing the characteristics of diets for different stages of larval development. If food rapidly passes through the gut, then supply of nutrients over a critical period of time has a better chance of being realized when a diet that has a comparatively low digestibility is used. The lack of high digestibility will be compensated by the Louis R. D'Abramo 148 comparatively large volume of food passing through the gut per unit of time. The comparatively long gut retention time of carnivorous fish and crustacean larvae presents a greater challenge because diets must be highly digestible when they pass through the gut.

Otherwise, adequate supply of essential nutrients for growth and survival per unit of time cannot be achieved. Achieving good digestibility of formulated diets for carnivorous fish and crustaceans is a difficult challenge because although sufficient enzyme activity is present in the primordial gut, the quality of enzymes is restricted. Ingredient sources of dietary nutrients must be compatible with the existing digestive capacity, particularly during early development of the gut. Larval diets cannot be simply ground pieces of a commercial diet that has proved successful for growth and survival of large specimens.

Bacteria

Previous focus on procedures to limit/remove the presence of bacteria from the culture water of a system where formulated feeds were being tested may, in fact, have been a bias that led to failure. Properly conditioned water with active bacterial populations, rather than "sterile" water, may be one of the more important contributors to successful larval culture using formulated diets. While recognizing the need to follow practices that will not induce abnormally high levels of bacteria, reduction to a level less than that found in the natural environment may reduce the probability of satisfactorily conditioning the gut for digestion of the microdiet. Indigenous bacterial flora may contribute significantly to larval digestion at certain stages of metamorphosis (Pinn, Rogerson & Atkinson, 1997). As gut retention time increase during larval ontogeny, the contributive role of microbiota to digestion may also increase. Microflora may serve as a supplementary source of food and microbial activity in the gut may be source of vitamins or essential amino acids (Dall & Moriarty, 1983). Also, establishment of the appropriate bacterial flora in the culture water may be conducive to the reduction of the incidence of Vibrio and the larval mortality that results.

The success with the feeding of some live foods may partly arise from the contribution of associated bacteria that is exposed to the gut epithelium of larval fish and crustaceans. The prevailing thought that bacterial levels must be all but eliminated because of potentially adverse effects on culture needs to be abandoned.

Culture Container

The design of the culture container can seemingly exert a marked effect upon the success of food acquisition, specifically for larval forms that are not filter feeding. Consumption of diet may not always be primarily determined by the chemical characteristics, physical properties or attractant value, but rather on the environment into which the formulated diet is introduced. For example, the shape of the container or the magnitude of the movement of water within the container may be important factors contributing to the success of larval culture. Larval forms of the American lobster are successfully grown in a cylindrically tapered tank that maintains suspension of both larvae and food, simulating the planktonic condition of the natural environment (Chang & Conklin, 1983).

Formulated diets for larvae 149 Provision of the proper contrast for larvae to locate food in the milieu of the tank has been assumed to the important for species of raptorial larvae. Provision of a light background colour is known to be a very important for some species of larval fish and crustaceans. Creating a color contrast for the efficient removal of uneaten feed from the bottom of a tank, may actually be counterproductive to strengthening food acquisition. Maintaining food in suspension for as long as possible would increase the frequency of encounter for those species of larvae that do not demonstrate a bottom feeding behaviour. Special light. Conditions may also be necessary to optimize food acquisition. Essentially, the culture conditions that afford larvae the best opportunity to encounter/locate and consume the formulated diet cannot be assumed to be similar.

MATERIALS AND METHODS

Relative growth, brood size and population dynamics, in *M. macrocopa* were studied in cultures raised in 5 litre glass troughs under laboratory conditions using different nutrient sources in condition viz., rice bran +cow dung, rice bran +poultry manure, rice bran +*chlorella*, rice bran and *chlorella*.

In the present studies extract of rice bran and cow-dung was used rice bran was micronized using a grinder and then was sieved to obtain the small particle easily consumable by *M. macrocopa*. Then the rice bran was soaked in water for about 1 hour and the suspended well mixed solution was added. The cow-dung was also powdered and soaked in water for 1-2 hour and then filtered through filter paper and then the filtrate was added to the stes.

In the set I, rice bran +cow dung were used in the ratio of 1:1 at the rate of 1 g l^{-1} each.

In the set II, rice bran +poultry manure were used in the ratio 1:1 at the rate of 1 gl^{-1} each.

In set III, rice bran +*chlorella* were used rice bran was used at the rate of 1 gl^{-1} and *chlorella* (cultured at laboratory lend) was added at the rate of 20 mg/l.

In set IV, rice bran alone was used at the rate of 2 gl^{-1}.

In set V, *chlorella* alone was used at the rate of 40 ml^{-1}/l.

The reading were done on alternate days. The sets were run in duplicate and water was changed twice a week, using sieves. The length and breadth were measured using ocular micrometer. Brood was counted unter microscope as the carapace is transparent. Population was counted with the help of dropper.

RESULTS AND DISCUSSION

Whether formulated diets that will completely substitute for live food can ever be developed remains a question. The complement of desired characteristics for success are often mutually exclusive. For example, creating a water stable particle that offers minimal leaching may be achieved at the expense of the loss of qualities of attraction, palatability or digestion. However, the leaching of water soluble nutrients does not present a conflicting problem if levels in excess of what is actually required are included to meet the requirements. Other problems may simply have to be confronted with an acceptance that the lack of mouthparts and the presence of a primordial gut in early larval stage of carnivorous species of marine fish require live food. A nutritionally complete diet that is stable, palatable, and easily digested can be developed with the understanding that the development of the gut in larvae is so dynamic that attempts at specialization will be futile. Future formulation and manufacture of larval diets must avoid becoming stage or species-specific and be based upon a general applicability. A good example is the recently developed egg yolk based diet for larvae of the freshwater shrimp *Macrobrachium rosenbergii* (Kovalenko et al., 2002). Fed exclusively for most of the larval cycle, growth and survival were at least 80% of that achieved with a live diet of newly hatched nauplii of Artemia. This diet is composed of readily available ingredients, is easy to make, and can be produced in a variety of forms, as a high moisture custard diet, dried to produced > 150 micron particles, or even spray-dried.

The ontogeny of larval nutrition is obviously complex and not well understood. Nonetheless, the development of formulated diets must be approached with a consciousness whereby biases and restrictive lofty goals may have to be surrendered and simplicity of thought is the guiding force. That is the challenge. Louis R. D'Abramo 150.

Effect of different nutrient media, viz., rice bran + cow dung, rice bran+ poultry manure, rice bran+*chlorella*, rice bran alone and *chlorella* alone, on growth pattern, brod size and population dynamics of *M. macrocopa*.

Table 1. Relative growth, brood size and population dynamics in *Moina macrocopa* under different nutrient media.

Sets		I. Rice Bran+Cow Dung				II. Rice+Poultry Manure				III. Rice Bran+Chlorella				IV. Rice Bran				V. Chlorella			
Days		Length (mm)	Breadth (mm)	Brood size	Pop.	Length (mm)	Breadth (mm)	Brood size	Pop.	Length (mm)	Breadth (mm)	Brood size	Pop.	Length (mm)	Breadth (mm)	Brood size	Pop.	Length (mm)	Breadth (mm)	Brood size	Pop.
Initial Inoculation		1.20	.99	7	20	1.20	.99	7	20	1.20	.99	7	20	1.20	.99	7	20	1.20	.99	7	20
5		1.00	.88	13	260	1.03	.84	10	140	1.01	.90	9	141	1.00	.90	15	200	1.01	.89	14	140
10		1.13	.90	14	20000	1.07	.99	16	1480	1.09	.97	11	1551	1.11	.97	17	1787	1.13	.98	15	1690
15		1.11	.96	16	4875	1.13	.98	15	4160	1.16	.98	13	4556	1.11	.97	16	6000	1.17	.98	15	5730
20		1.19	.99	18	8635	1.11	.99	17	7386	1.18	1.00	19	7229	1.21	1.00	20	9600	1.16	.99	18	7563
25		1.23	1.29	20	12590	1.18	1.00	17	11900	1.21	1.01	18	12175	1.28	1.00	21	14000	1.20	.98	10	1210
30		1.34	1.03	13	10612	1.28	.88	11	9645	1.29	.89	12	9952	1.30	.90	13	10800	1.30	.92	10	10031
35		1.24	1.00	20	11600	1.20	1.03	20	10205	1.22	1.00	18	10887	1.24	1.00	20	11349	1.20	.89	15	11450
40		1.25	1.02	21	12769	1.22	1.00	19	11700	1.20	1.09	18	11230	1.20	1.10	21	11700	1.81	.90	14	11889
45		1.28	1.19	22	13255	1.28	1.10	20	12568	1.27	1.10	21	12180	1.35	1.00	20	13170	1.30	.91	14	11459
50		1.28	1.12	18	14500	1.21	1.00	16	13750	1.20	1.09	18	1200	1.29	.96	14	11800	1.21	.89	12	10350
55		1.42	1.10	20	12860	1.30	1.00	18	11380	1.25	.97	17	10830	1.20	.97	14	10230	1.01	.82	6	9000
90		1.29	1.00	15	92550	1.21	.84	11	858	1.19	.98	12	8242	1.09	.96	12	9020	1.00	.82	5	3500s

Growth Patterns

During present studies different nutrient sources viz., rice bran +cow dung, rice bran + poultry manure, rice bran + *Chlorella*, rice bran alone and *Chlorella* alone, in five sets of different combinations have been used (as culture media) for the culture of *Moina macrocopa* in the laboratory. (Figs. 1-2).

Length of *M. macrocopa* a Growth Parameter

A look at the Table 1; Fig. 1 indicates that the mean length of *M. macrocopa* in all the combinations of culture media records of steady growth after 5th day from the day of initial inoculation of gravid adults followed by sudden decline after 30th day. These results are in agreement with those of Langer (1987 & 1991 unpublished), who also observed similar decline in average mean length due to appearance of young individuals. Similar type of trend in decline in average body length has also been reported by Jana and Pal (1985a) in another species *Moina micura* where authors (*op. cit.*) have attributed such decline in the body length with increase in the population density of individuals.

The maximum mean length in *M. macrocopa* is observed in set (Table 1; Fig.1) fed on rice bran +cow dung (1.42 mm), followed by rice bran (1.35 mm) and rice bran + poultry manure (1.30 mm), *Chlorella* (1.30 mm) and rice bran + cow dung gives best results in linear growth of *Moina macrocopa* in the laboratory.

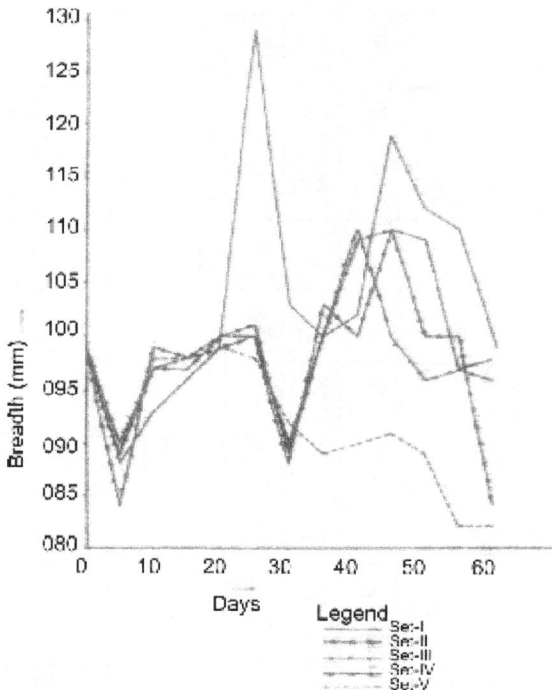

FIG. 1.   Graph showing comparative study of Breadth with No. of Days under different nutrient media in the culture studies of *Moina macrocopa*.

Brood Size

The present observations as indicated in Table 1; Fig. 2 reveal that the brood size in *M. macrocopa* recorded a steep increases on the 25th followed by a second peak on 35th day and third peak on 55th day, in all the sets except for the one cultured on *Chorella* where the egg production recorded a bimodal pattern instead of trimodal pattern in rest of the cases. A bimoda pattern in egg production is a known phenomenon in *Daphnia carinata* (Jana and Pal, 1985a) where mahua oil cake was used as food. The maximum number of eggs (22) reported in culture raised on rice bran+cow dung, supports the view that this nutrient combination is best to provide enough energy for growth and reproduction of *M. macrocopa* and may be due to the assimilation of excess of energy form this nutrient combination which is better directed towards egg production. Similar conclusions have been arrived at by Crabtree (1975) and Paloheimo *et al.* (1982), in case of *Daphnia*, where it is suggested that, allocates an increasing fraction of assimilable energy (in excess to maintenance requirement) to reproduction.

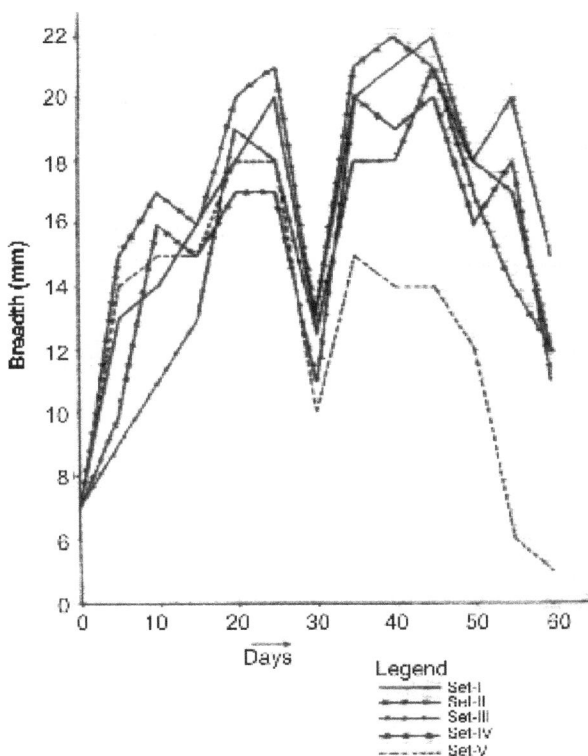

FIG. 2.   Graph showing comparative study of Brood Size with No. of Days under different nutrient media in the culture studies of *Moina Macrocopa*.

The decline in egg production observed presently (Table 1; Fig. 3) may be attributed to the age of culture as earlier suggested by De Pauw *et al.* (1981), who opined that old cultures develop unfavourable conditions due to sedimentation of unwanted toxic metabolites and that when population increase, competition for food also increase as has been observed *in Daphnia* where the adults produce no eggs at all (McCauley *et al.* (1990), conversely, De Pauw *et al.* (1981) also reported that when *D. magna* is raised on rice bran, the lowering of population density by harvesting decreased competition percentage of ovigerous females in the population, hence results in production of greater number of eggs. These results indicate an inverse relationship of egg production with competition.

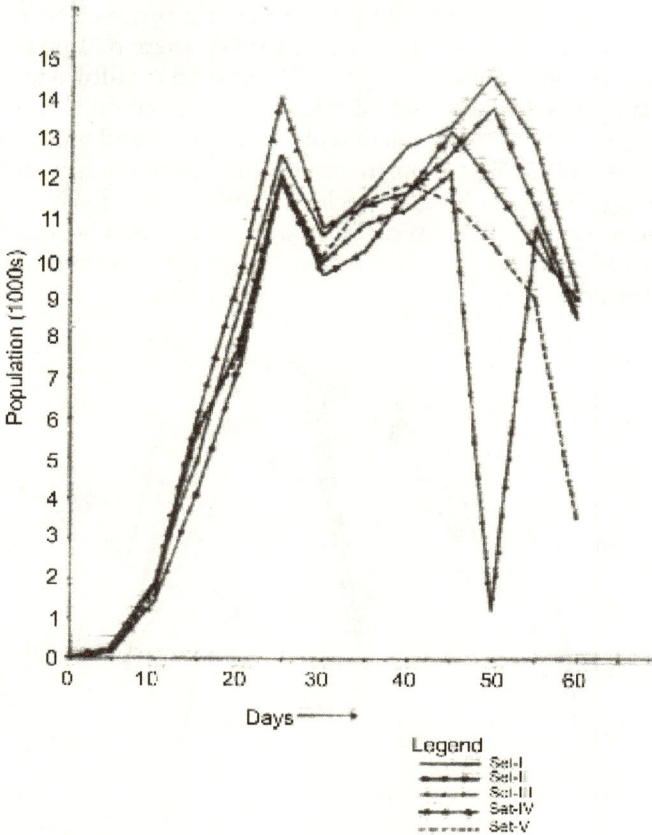

FIG. 3.　Graph showing comparative study of Brood Size with No. of Days under different nutrient media in the culture studies of *Moina Macrocopa*.

Population size

The population size varied markedly in all the five different nutrient sources presently experimented upon. The maximum average number of *Moina macrocopa* individuals have been obtained in case of rice bran+ cow dung combination (14500 inds/2 lit) and minimum in *Chlorella* (12100 inds/2 lit).

The first peak in population size has been registered on the 25th day in all the nutrient combinations used (Table 1; Fig. 18), followed by a sudden decline in population size on 50th attributed such population fluctuations in number to carrying capacity of the medium followed by the internal adjustments by developing organisms to obtain fitness of the population. Odum (1971) also concluded that there is a strong tendency for all populations to evolve through natural selection towards self regulations and stability.

De Pauw *et al.* (1981) have indicated better results when cultures of fish food organisms were raised with rice bran alone where conversion efficiency is of the order of 1 to 1, and is reported to be an excellent medium as compared with other feeds (such as manures).

Thus from the present studies it is indicated that the best combination for culturing *M. macrocopa*, out of all the combination of nutrient sources presently tried is rice bran +cow dung, which suggests that these two nutrient sources when applied in combination yields better results for production of these fish food organisms.

SUMMARY AND CONCLUSIONS

Effect of different nutrient media on growth patterns, brood size and population dynamics of *M. macrocopa*:

Growth patterns

During present studies on *M. macrocopa* a steady growth has been observed to be followed by a sudden decline.

The maximum growth (mean length of *M. macrocopa*) has been observed in set fed on rice bran+cow-dung (1.42 mm).

It is thus concluded, from the data on growth, from all the nutrient sources tried presently, the combination of rice bran+cow–dung is the best so far as the linear growth of *M. macrocopa* is concerned.

Brood Size

Egg production in *M. macrocopa* recorded a trimodal pattern in all the nutrient combination, except for *Chlorella*, where is remained bimodal.

Maximum average number of eggs (22) have been observed in culture raised on rice bran +cow-dung which provide enough energy for growth and reproducing in *M. macrocopa*.

It is indicated that age of culture also effects the egg production in *M. macrocopa* (wherein old cultures have been observed to be unfavourable for egg production).

Population size

Maximum average number of *M. macrocopa* individuals have been obtained in rice bran +cow-dung combination (14500 inds/2 lit.). It is thus concluded that the best combination, for culture of *M. macrocopa* (out of all the combinations of nutrient sources presently tried) is rice bran + cow-dung, which when applied in combination yiel better results.

Effect of inoculums density on growth patterns, fecundity and population dynamics of *M. macrocopa*:

Body length /body breadth

Body length/body breadth recorded variations in *M. macrocopa* with varying inoculum density but no significant relationship between inoculums density and growth (body length/ body breadth) has been observed.

Population size

Heavy mortality among population of M. *macrocopa* has been observed in all the experimental sets due to increased competition for food and space among individuals. Minimal space appears to be a critical factor for such heavy mortality.

Brood size

Brood size in M. *macrocopa* do not show any specific variation in relation to the different inoculums densities but overall fecundity increase when grown under low inoculum density (1 ind/200 cc) in the laboratory.

REFERENCES

Bautista M. N., Millamena O. M., Kanazawa, A., (1989). Use of kappa-carrageenan microbound diet (CMBD) as feed for *Penaeus japonicus* larvae. Marine Biology 103, 169-173.

Blair T., Castell J., Neil S., D'Abramo L., Cahu C., Harmon P., Ogunmoye, K., (2002). Evaluation of microdiets versus live feeds on growth, survival and fatty acid composition of larval haddock *(Melangrammus aeglefinus)*. Proceedings of the Nutrition and Feeding in Fish Symposium, in press.

Cahu C. L., Zambonino Infante, J. L., (1997). Is the digestive capacity of marine fish larvae sufficient for compound diet feeding? Aquaculture International 5, 151-160.

Chang E. S., Conklin, D. E., (1983). Lobster (*Homarus*) hatchery techniques. In: Handbook of Mariculture, Vol. 1 Crustacean Aquaculture (ed. By J. P.   Mc Vey), pp. 271-275. CRC Press, Boca Raton, FL, USA.

Clawson, J. A., Lowell, R. T., (1992). Improvement of nutritional value of Artemia for hybrid striped bass/white bass (*Morone saxatilis X M. chrysops*) larvae by n-3 HUFA enrichment of nauplii with menhaden oil. Aquaculture 108, 125-134.

Crabtree, S.J. (1975). Population dynamics in Daphnia pulex, Dissertation. University of Toronto, Toronto, Ontario, Canada.

D'Abramo, L. R., (1979). Dietary fatty acid and temperature effects on the productivity of the cladoceran *Moina macrocopa*. Biological Bulletin 157, 234-248.

Dall, W., Moriarty, D. J. W., (1983). Functional aspects of nutrition and digestion. In: The Biology of Crustacea, Volume 5, Internal anatomy and physiological regulation (ed. By L. H. Mantel) Academic Press.

De Pauw, Laureys, P. and Morales, J. (1981). Mass cultivation of Daphnia magna straus on rice bran. Aquaculture, 25 (2,3): 141-152.

De Pauw, M., De Leenhear, L. Jr., Laureys, P., Morales, J. and Reartes, J.(1980). Culture D'A 1 gues et d' Invertebres sur Dehets Agricoles. In. R. Billard (Ed.), La Pisciculture en Etang, INRA, Publ. Paris: 189-214.

Fernandez-Diaz C., Pascual, E., Yufera, M., (1994). Feeding behavior and prey selection size of gilthead sea bream, Sparus aurata, larvae fed on inertand live food. Marine Biology 118, 323-328.

Garcia-Ortega, A., Verreth, J., Coutteau, P., Segner, H., Huisman, E. A., Sorgeloos P., (1998). Bichemical and enzymatic characterization of decapsulated cysts and nauplii of the brine shrimp Artemia at different development stages. Aquaculture 161, 501-514.

Chang E. S., Conklin, D. E., (1983). Lobster (*Homarus*) hatchery techniques. In: Handbook of Mariculture, Vol. 1 Crustacean Aquaculture (ed. By J.P. Mc Vey), pp. 271-275. CRC Press, Boca Raton, FL, USA.

Clawson, J. A., Lowell, R. T., (1992). Improvement of nutritional value of Artemia for hybrid striped bass/white bass (*Morone saxatilis X M. chrysops*) larvae by n-3 HUFA enrichment of nauplii with menhaden oil. Aquaculture 108, 125-134.

Crabtree, S. J. (1975). Population dynamics in Daphnia pulex, Dissertation. University of Toronto, Toronto, Ontario, Canada.

D'Abramo, L. R., (1979). Dietary fatty acid and temperature effects on the productivity of the cladoceran *Moina macrocopa*. Biological Bulletin 157, 234-248.

Dall, W., Moriarty, D. J. W., (1983). Functional aspects of nutrition and digestion . In: The Biology of Crustacea, Volume 5, Internal anatomy and physiological regulation (ed. By L.H. Mantel) Academic Press.

De Pauw, Laureys, P. and Morales, J. (1981). Mass cultivation of *Daphnia magna* straus on rice bran. Aquaculture, 25 (2,3) : 141-152.

De Pauw, M., De Leenhear, L. Jr., Laureys, P., Morales, J. and Reartes, J. (1980). Culture D'A1 gues et d' Invertebres sur Dehets Agrocoles. In. R. Billard (Ed.), La Pisciculture en Etang, INRA, Publ. Paris: 189-214.

Fernzndez-Diaz C., Pascual, E., Yufera, M., (1994). Feeding behavior and prey selection size of gilthead sea bream, *Sparus aurata*, larvae fed on inert and live food. Marine Biology 118, 323-328.

Garcia-Ortega, A., Verreth, J., Coutteau, P., Segner, H., Huisman, E. A., Sorgelloos P., (1998). Biochemical and enzymatic characterization of decapsulated cysts and nauplii of the brine shrimp Artemia at different developmental stages. Aquaculture 161, 501-514.

Jana B. B and Pal, G. P. (1985a). Relative growth and egg production in Daphnia carinata under different culturing media. Limnologia and (Berlin), 16 (2) : 325-339.

Jana B. B and Pal, G. P. (1991). Culture and some biological aspects of some important fish food organisms Ph. D. Thesis submitted to the University of Jammu. (1985b). Effects of inoculums density on growth, reproductive potential and population size in Moina *micrura* (Kurz.). Limnologica (Berlin), 16(2) : 315-324.

Jones, D.A., Kanazawa, A., Rahman, S. A., (1979). Studies on the presentation of artificial diets for rearing the larvae of *Penaeus japonicus* Bate. Aquaculture 17, 33-43.

Jones, D.A., Yule , A. B., Holland, D. L., (1997). Larval nutrition. In: Crustacean Nutrition (ed. By L. R. D'Abramo, D. E. Conklin & D. M. Akiyama), World Aquaculture Society.

Kamarudin, M. S., Jones, D. A., Vay, L., Abidin, A. Z., 1994. Ontogenetic changes in digestive enzyme activity during larval development of *Macrobrachium rosenbergii*. Aquaculture 123, 323-333.

Kolkovski, S., Tandler, A., Izquierdo, M. S., (1997). Effects of live food and dietary digestive enzymes on the efficiency of microdiets for seabass (*Dicentrarchus labrax*) larva. Aquaculture 148, 313-332.

Kolkavski, S., Tandler, A., Kissil, G. W., Gertler, A., (1993). The effect of dietary exogenous digestive enzymes on ingestion, assimilation, growth, and survival of gilthead seabream (*Sparus aurata*, Sparidae, Linneaus) larvae. Fish Physiology and Biochemistry 12, 203-209.

Koshio, S., Kanazawa, A., Teshima, S., Castell, J. D., (1989). Nutritive evaluation of crab protein for larval *Penaeus japonicus* fed microparticulate diets. Aquaculture 81, 145-154.

Kovalenko, E. E., D'Abramo, L. R., Ohs, C. L., Buddington, R. K., (2002). A successful microbound diet for the larval culture of freshwater prawn (*Macrobrachium rosenbergii*). Aquaculture 210, 385-395.

Kumlu, M., Jones, D. A., (1995). The effect of live and artificial diets on growth, survival, and trypsin activity in larvae of *Penaeus indicus*. Journal of the World Aquaculture Society 26(4), 406-415.

Kurmaly, K., Jones D. A., Yule, A. B., (1990). Acceptability and digestion of diets fed to larval stages of *Homarus gammarus* and their role of dietary conditioning behaviour. Marine Biology 106, 181-190.

Langer, S. (1987). Culture and some biological aspects of a cladoceran, *Simocephalus vetulus*, H. Phil. Dissertation submitted to University of Jammu. p. 115.

Langer, S., (1991). Culture and some biological aspects of some important fish food organisms Ph. D. Thesis submitted to the University of Jammu.

Lovett, D. L., Felder, D. L., (1989). Ontogeny of gut morphology in the white shrimp *Penaeus setiferus* (Decapoda, Penaeidae). Journal of Morphology 201, 253-272.

Lovett, D. L., Felder, D. L., (1990). Ontogenetic change in digestive enzyme activity of larval and postlarval white shrimp *Penaeus setiferus* (Crustacea, Decapoda, Penaeidae). Biological Bulletin 178, 144-159.

Mc Cauley, E., Murdoch, W. W., Nisbet R. M. and Gurney W. S. C. (1990). The physiological ecology of Daphnia: Development of a Model of Growth and Reproduction. Ecology, 71 (2): 703-715.

Munilla-Moran, R., Stark, J. R., Barbour, A., (1990). The role of exogenous enzymes on the digestion of the cultured turbot larvae, *Scophthalmus maximus* L. Aquaculture 88, 337-350.

Odum, E. P. (1971). Fundamentals of Ecology, 3rd ed., W. B. Saurders Co., Philadelphia, London, Toronto.

Ohs, C. L., D'Abramo, L. R., Buddington, R. K., Robinette, H. R., Roethke, J. M., (1998). Evaluation of a sprydried artificial diet for larval culture of freshwater prawn *Macrobrachium rosenbergii*, and striped bass, *Morone saxatilis*. Aquaculture Nutrition 4, 73-82.

Paloheimo J. E., Crabrtee S. J. and Taylor WID. (1982). Growth model of Daphnia. Canadian Journal of fisheries and Aquatic Sciences, 39 : 598-606.

Pinn, E. H., Rogerson, A., Atkinson, R. J.A., (1997). Microbial flora associated with the digestive system of *Upogebia stellata* (Crustacea: Decapoda: Thalassinidea). Journal of Marine Biology Association UK 77(4), 1038-1096 (cited from an abstract).

Rabe, J., Brown, J. A., (2000). A pulse feeding strategy for rearing larval fish: an experiment with yellowtail flounder. Aquaculture 191, 289-302.

Teshima, S., Ishikawa, M., Koshio, S., (1993). Recent developments in nutrition and microparticulate diets of larval prawns. The Israeli Journal of Aquaculture- Bamidgeh 45(4), 175-184.

Teshima, S., Kanazawa, A., (1983). Effects of several factors on growth and survival of the prawn larvae reared with microparticulate diets. Bulletin of the Japanese Society of Scientific Fisheries 49, 1893-1896.

Tucker, J. W. Jr., (1998). Marine fish culture. Kluwer Academic Publishers, Boston, 750 p.

Zambonino Infante, J. L., Cahu, C. L., (1994a). Development and response to a diet change of some digestive enzymes in sea bass (*Dicentrarchus labrax*) larvae. Fish Physiology and Biochemistry 12, 399-408.

Zombonino Infante, J. L., Cahu, C. L., (1994b). Influence of diet on pepsin and some pancreatic enzymes in sea bass (*Dicentrarchus labrax*) larvae. Comparative Biochemistry and Physiology 109A, 209-212.

ロロロ

7

FISH DIVERSITY IN A LOTIC ENVIRONMENT, MUNAWAR TAWI OF RAJOURI, J&K STATE

A.K. Verma, Raheela Mushtaq and Anuradha Gupta

ABSTRACT

India is one of the 18 mega-diversity nations and displays significant animal diversity. Indian fauna constitutes 7.28% of the total species found in the world and the endemic species are 9.23% of the total endemic species in the world. India is a house to about 11.7% of all fish fauna (Belsare, 2007). Some of the largest rivers in the world are in India and 80% of the total length is covered by the 14 major rivers which are rich in fish resources. The range and abundance of the species change dramatically due to the changed climatic conditions and those species which cannot adapt to the changed climate may migrate to the new locations. Conservation of biological diversity is essential for the survival of the human race, because it not only provides livelihood to them but also mitigates climatic changes and also maintains samples of unchanged biotic communities in their natural form for breeding and study purposes. Earlier, biodiversity conservation was limited to saving genes, species and habitats, but now a new conservation philosophy based on saving biodiversity, has emerged. Species-focused conservation efforts are, therefore, essential together with efforts to conserve habitat & the ecosystems.

Thus, there is an urgent need for a comprehensive inventory and status survey of taxonomic groups especially in the case of fish diversity. Certain areas like the Himalayas (J&K included), Andaman and Nicobar Islands are still partially surveyed. There is, therefore an urgent need for periodic review of research programme from time to time in order to understand structure (diversity and composition), function & interaction amongst and within an ecosystem. The methods to understand ecosystem services and impacts of different resource uses in various environments should be developed and standardized. Thus the fresh water fish diversity can serve as a plethora of livelihood and biodiversity of conservation values, besides serving as source of protein and food for the increasing human population. The extensive farming of fish is also more sustainable for livelihood of the rural poor because, it serves as security of food, health and earned income. (Belsare 2005).

Munawar Tawi, a tributary of River Chenab flowing through district Rajouri, the present study area, provides a suitable habit for fish and also acts as a resort for a variety of birds for foraging. This water body also harbors a vast array of other fauna besides foraging birds. But hardly any account is available on the ecology or biodiversity of this river. Despite of the reason that this water body runs along the indo-Pak border *i.e.* Rajouri- Nowshera sector of J&K state and there is a vast potential of fisheries in this belt of the Pir – Panjal mountain range. Through fishing practices dual purpose viz. employment generation and conversation measures for future generation can be realised. Since the present study has revealed the occurrence of both the palearctic and oriental influence on the fish of this region, it has been related to the geographic history of India and the conversation of rare and endangered species has become an important priority and tool for conservation of overall biological diversity. It is in this backdrop that present investigation was planned and initiated.

There is also dire need of humans to connect with nature through biodiversity conservation. The present work was, therefore, undertaken to document the fish diversity in this water body. The data on the history of occurrence/distribution of the fish was cross checked with the local people. From this study it could be stated that this river houses a wide variety of faunal elements and further more detailed studies of the biota is necessary to recommend conservation measures.

Keywords: *Pir-Panjal range, Rajouri river, Fish diversity, Conservation, Munawar Tawi.*

INTRODUCTION

India with its varied physical and the meteorological features is a vast sub-continent singularly blessed with unparalleled aquatic resources harbouring one of the richest wealth in the world. The hill streams with torrential falls over precipitous rocks, the broad and long placid rivers, lakes, backwaters and estuarine systems form the major inland aquatic media accommodating diverse animal groups. Nowhere else does such a comparable miniature world of unresolved mysteries to be probed exist as in any freshwater habitat be a lake or a river. The hill-streams though posses a unique fish-fauna of their own, but they have a different kind of problem because of their terrain, altitude and physiography.

Our country has neglected the great contributions which freshwaters can provide in the form of food fishes. In most parts of the world freshwater fishes are consumed in large quantities and reared artificially on a large scale to provide food to ever increasing human population. Scientific studies have confirmed that fish is highly nutritive and rich source of proteins, vitamins and minerals for which there is a great demand in all the countries of the world.

In terms of food value, fish with its high contents of proteins, fats and rich source of vitamins & minerals is often referred to as rich food for poor people (Bene and Heck, 2005). Fish contains all the eight essential amino-acids and rich source of vitamins A, B and D. the minerals present in fish include iron, calcium, zinc, selenium phosphorous and fluorine which are highly 'bio-available' and hence they are easily absorbed by the body. Development of fishery therefore, shall be not only a supplement to our diet but it may also increase the country's economy by attracting the tourists. The rationale of focusing on the fish diversity is based on its extraordinary potential relating to the human-health and excellent food value. Fish also forms an important link in the food chain operating in an ecosystem through the growth of planktons and the benthos.

Studies on fish diversity of the J&K state attracted the attention of several investigators in the past who confined their studies mostly to the fishes of Kashmir valley leaving the other two regions (Jammu and Ladakh provinces) of the state poorly surveyed or unexplored Das and Nath, 1996, Sharma, 1978) till the first comprehensive report on the ichthyo-fauna of the state reporting 125 species of freshwater fish representing the 18 families and 51 genera was published (Nath, 1986).

The faunal elements in general, and the fish in particular, inhabiting the hill-streams of Jammu province are poorly known and hence the present study has been initiated to explore the fish diversity (piscine wealth) in a lotic environment i.e. Munawar Tawi, a tributary of Chenab river flowing through the border district Rajouri of J&K state. It will be interesting to evaluated what is the real present day status of the fish diversity in aquatic network of this hill stream flowing through the pir- Panjal belt of the Himalayan ranges which would offer a great scope to further explore its hydrological characteristics in future.

MATERIALS AND METHODS

Study Area

The state of Jammu and Kashmir stretches between 32°26′ E to 80°30′ E longitudes. The state morphologically is divisible into three distinct regions *viz.* Jammu, Kashmir and Ladakh, each region having distinct physiographic features and ichthyofaunal characterictics. The state is drained by the mighty Indus, Jhelum, Kishen Ganga and Chenab rivers and their tributaries. Out of these the Chenab river rises well to the north of the greater Himalayas and pierces through the main ranges of the Himalayas. Chenab river drains the southern sides of the Pir-Panjal which forms the greater part of Jammu division.

Chenab river is actually formed by the confluence of the Chandra and Bhaga streams rising in Lahul and Spiti regions of the Himachal Pradesh. Both these streams have their origin from glaciers that slide down from slopes of the Baralacha peak at an altitude of 5000 m.s.l. This river covers a distance of 200 km in the valley of the Himachal Pradesh before entering the J&K state near Padder, Kishtwar and after passing through Doda, Ramban, Reasi and Akhnoor at Jammu it ultimately enters Pakistan. Some of its tributaties in higher reaches are torrential, snow fed and hence cold water streams. The total length of the Chenab river is approx 960 kms.

The Rajouri district (30°40′ N and 74°31′ E) is an area of diverse ecologies. While western and southern parts are warm temperate the northern and eastern parts are cold temperate having conditions similar to that of Kashmir. Munawar Tawi also called Rajouri River originates in Thanna mandi area in the north of Rajouri and after flowing through Rajouri, Chingus and Nowshera, it enters Pakistan at Munawar (Punjab) and near Marala it joins the Chenab River. Various contributaries to Munawar Tawi includes Darhali Nallah, B.G. Nallah, Karsi-Nallah, Jamir Nallah and Chingus Nallah.

METHODOLOGY

The field work was carried out during the years 2010-11 and collection sites visited fortnightly excepting flash-floods and inclement weather when collection was not possible especially during monsoons and severe winters. In view of the expected wide diversity of freshwater fishes in different regions of the Jammu province, the method of collection varied from site to site and species. While in head region of the stream (Munawar Tawi), fishes were collected mostly by damming up portions of stream and collecting the exposed specimens by means of a dip net. Dredging with a copper wire dredge was resorted for the species which hide under and among the submerged stones and pebbles. Fish inhabiting pool and riffle section of stream were caught using a hook and wire with a live bait. The collected specimens were transported to the laboratory in polythene bags for morphometric studies. Identification was done after following the keys by Nath (1986), Jayaram, (1986, 1999) and subsequently preserved in formal-alchol (3:7) mixture.

OBSERVATIONS

Various fishes recovered and the key characteristics taken into consideration while identifying fish species collected from the study area have been summarized as:

Taxonomic description:

Genus: *Mastacembulus*

Order Mastacembliformis

Family Mastacembelidae

Pre-orbital spine present and caudal fin confluent with dorsal fin and anal fin in case of *M. armatus*.

Vernacular name: groach

Genus: *Glyptothorax*

Order Cyprinifformes

Sub-order Siluiroidei (body without scales)

Family: Sisroidae

Head and body ventrally flatted, fins horizontal, eyes reduced and dorsally located in case of *G.pectinopterus*.

Vernacular name: kingad.

Genus **Mystus**

Order: Cypriniformes

Sub order Silburoidei

Family: Bagaridae

Vernacular name: palu

Body not ventrally flattened, fins not horizontal, eyes, laterally located in case of *M.bleekeri*.

Genus: *Ompok*

Order: Cypriniformes

Suborder Siluiodiae

Barbles four, cleft of mouth never extends beyond the eye orbit in case of *O.bimaculatus*.

Vernacular name: mali

Order: Cyprininoidei

Family:Cyprinidae

Barbles may or may not be present, when present not more than 4; scales moderate or large. Mouth anterio, dorsal fin with a spine, jaws not compressed and anal fin short. This family was represent by the following genera in Munawar Tawi:-

Genus: **Tor**

Mouth with thick lips

Species: *T. tor*

Vernacular name: lachu

Species: *T.putitora*

Vernacular name: kheri

Genus: *Garra*

Body ventrally flattened, and dorsal fin without a spine

Species: *G. gotyla*

Vernacular name: moda

Genus: *Schizothorax*

Body not ventrally flattened, dorsal fin with a spine

Species: *S.richardsoni*

Vernacular name:las

Genus:*Labeo*

Sucker absent, lips thick and continuous at the junction of jaw angles.

Species:*L.dero*

Vernacular name: akhrote

Genus: *Puntius*

Mouth without thick lips and barbles are absent. A black spot at the base of dorsal fin is present.

Species: *P. sophore*

Vernacular name: chapta

Besides the fishes the decapods crab, pigmy back swimmer, odonates Segmentina mollusc, frogs and tadpoles of frog were also recovered from this region.

Genus: **Rasbora**

Body bilaterally compressed dorsal fin, single, situated near to the base of the caudal fin, body is silver coloured; tail homocercal, lateral line present.

Vernacular name: sunehri.

DISCUSSION

Out of existing 59,811 species of vertebrates, fish contribute about 30,000 living species and occupy almost all conceivable aquatic habitats (IUCN, 2007). In India, there are about 2500 species of fishes, of which 930 live in freshwaters and 1570 in sea & estuarine waters (Kar, 2003). The great fish diversity of this region resulted in the north-eastern India being identified as a biodiversity 'hot spot' by the world Conservation Monitoring centre (WCMC, 1998) and consequently attracted the attention of many biologists from India & abroad. It was also been stated that for proper and systematic genetic study of a species (fish), it is necessary to have the basic information regarding its diversity, conservation status and the economic importance of the species (Kamal and Parvez, 2011).

Taxonomic surveys conducted during almost last two centuries on fish diversity by various groups and authors have revealed presence of 256 species of freshwater fishes belonging to 65 genera, 12 families and 5 orders have been reported from various aquatic resources of the Western to the north-central Himalayan region. It has also been reported that the freshwater fish fauna of the Indian sub-continent has elements in it, which are common to the China and Indo-Malayan sub-region (Jayaram, 1981).

Regarding the occurrence and assemblance of freshwater ichthyofauna of India in relation to adjoining countries, Hora (1937) advocated a south-east Asian origin of fish fauna of India and concluded that not only had the Indian fish fauna a marked Malayan affinity but was also related to the fauna of Thailand, southern China and Cochin-China regions. According to Hora's View, the ichthyofauna spread westwards in successive waves of migration, first to India, and later on to the Africa, at a time when the two landmasses were connected. Most of the present day fish fauna of India has descended from he Pleistocence migrants while, the migration along the Himalayas westwards commenced in the Pliocence and continued uninterruptedly till the late Pleistocence, when the reversal of the Himalayas had occurred entirely during the Pleistocence.

Silas (1960) described that the bulk of indigenous fish fauna of Kashmir valley is composed of the palearctic elements of central Asiatic stock, which during the second interglacial period migrated south-west wards through Kashmir but was stopped from entering the India region by the closed mountain ranges coinciding with the present Pir-Panjal ranges. Subsequent invasions have occurred during the interglacial periods and flash floods, the last being about 25,000 years ago.

The origin of fishes of Jammu province is from the Indo Gangetic region to which region there was migration from East to West from South-East Asia. Evidence exists that the fishes of Jammu region have probably migrated to his present day habit at only in late Pleistocene (Das and Nath, 1966). It is significant to note that as many as 56 species of river Kosi drainage system as reported by (Khan and Kamal, 1979) occur in the fish fauna of Jammu region. Moreover, out of 31 species of freshwater fishes recorded from river Alkananda (Badola and Singh, 1981) as many as 17 species also occur in Jammu province.

The Poonch-Rajouri region of Jammu Province is, thus the only region of the state which exhibits a rare combination of palaearctic fish species with oriental and the Indo-Malayan fishes. This is also true about several other vertebrates and invertebrates (Mammals, Reptitles, Amphbians, Molluscs & Crustaceans etc.) as explained by Nath (1986). This indicates that Jammu province of J&K state has served as an aqueduct for the passage of fishes and also a number of other animals east to west. A detailed taxonomic study of fish specimen collected from Munawar Tawi, Rajouri reveals that many of them differ from the holotypes in several important details. The occurrence fo *Schizothorax, Tor, Labeo* and *Garra* of Kashmir region in the Rajouri River is not surprising as all these fishes possesses adequate adaptive modifications to allow them to subsist in torrential streams of Rajouri-Poonch region. Prolonged evolution under severe mountain terrain conditions resulted in the development of special adaptive mechanisms and subsequent modifications like reduction of scales, number of pharyngeal teeth, barbles, reduced or depressed body etc. in these torrential stream fishes. The morphological adaptations as recorded in the hill-stream fishes enable them to stay without being washed away by the powerful water currents flowing over and around their bodies.

When we compare the 91 species of freshwater fishes of Jammu region with 36 species of Kashmir region, it is apparent that there is a paucity of endemic fish species in Jammu region. It is due to the complete isolation of Kashmir form Jammu region by the Pir-Panjal mountain barrier which might have excluded many of the fish species of Kashmir region from Jammu and vice-versa. That a few species of Kashmir region like, *Schizothorax richardsoni, Garra gotyla, Labeo sps and Glyptothorax* speices which have also been reported in present study indicates that these species have established themselves in the formidable mountain barriers in Jammu region affords another evidence of the great climatic and ecological tolerance exhibited by these fishes which probably established together in both the regions in the late Pleistocene, when the present lofty Pir-Panjal range had not risen so high.

Schizothorax richardsonii and *Nemacheilus* have rarer distinction of having a much wider distribution, being also found in the Ladakh region, lower altitudes of the Kashmir valley as well as Rajouri, Poonch, Doda and Bhaderwah region of Jammu province. Likewise, complete absence of the representatives of the family bagaridae, heteropneustidae and mastecemblidae, reported from Munawar Tawi and other water bodies of Jammu region, from Kashmir region indicates that the high Pir- Panjal range was already established when these fishes migrated towards the Kashmir valley. Lastly the occurrence of fish species, *Tor* in the Rajouri River is another salient feature of the fish diversity of this region as this fish was previously known from range of southern India only. Apparently, these species have a much wider range of distribution than was previously known and further extends to northern India as well.

While adaptation and diversification, over the past billions of years tend to increase the number of species, the increasing human population and its interference into the natural habitat and rapid industrialization, urbanization, continuously play a major role in the decline of species diversity. Further more, in the tropical developing countries the process of species extinction and genetic loss are very common due to destruction of aquatic ecosystems, irrational harvesting of juveniles, addition of the pollutants into the aquatic habitats and genome homogenization, which plays an important role towards fish population decline.

NEED FOR CONSERVATION

The declining fish species and their populations necessitate robust conservation strategies which include all the necessary measures for protecting them. These strategies will ensure sustainability and diversity of species which shall help to improve the livelihood of maintaining minimal viable population of rare and endangered species. Cataloging fish diversity is important to identify which fish species are critical to sustainability of an aquatic ecosystem. Fish conservation can be done either *in-situ* (natural habitat) or *ex-situ* (cryo preservation of genes). Hence the pre-requisite for species conservation is development of the data-base on the species diversity. The genetic characterization and gene banking (cryo preservation) which wil be a subsequent step towards further conservation of the species at the molecular level.

The dwindling population of variety of living forms especially the fish fauna has been the area of great concern amongst the scientific community. A comprehensive remedial plan to check this alarming decline in biodiversity is the need of the hour. The particular emphasis on studies on Indian fish fauna is largely due to the fact that Indian sub-continent is the home for great biodiversity which is being systematically eroded by vested interests and unscrupulous lobbies. Hence, it greatly underlines the need for a passionate movement to conserve our legendary legacy before it becomes too late. Admittedly, the theme is very vast, however, this write-up should serve as an initiative for those wishing to pursue a fascinating theme in future.

ACKNOWLEDGEMENTS

We express our deep sense of gratitude to the local people of the study area who helped us in furnishing all the necessary information in the vernacular names of the fishes available in their nearest site. Thanks are also due to the several key informers & fishermen of the study site for their active support in the field work.

REFERENCES

Badola, S. P. and Singh, H. R., (1981). Fish and Fisheries of the River Alaknanda *Proc. Nat. acad. Sci. India*. B 51(1):133-142.

Bene, C. and Heck, S., (2005). Fish and Food Security In Africa. *NAGA World Fish Quaterly*, 28:8-13.

Das, S. M. & Nath, S., (1996). The Ichthyofauna of Jammu province, India. *Kashmir sci.* 3(1-2):65-78.

Hora, S. L., (1937). Geographical Distribution of Indian Freshwater fishes and its bearing on the probable and connection between India & adjoining countries. *Curr: sci*, 7:351-356.

IUCN (2007). IUCN red list of threatened speices. www.iucnredlist.org j.hum. gen.4:397-401.

Jayaram, K.C. (1999). Handbook of the freshwater fishes of india & oriental region. *NPH publ. house, N. Delhi*, pp. 998

Jayaram, K.C., (1981). Handbook of the Freshwater Fishes of India, Pakistan, Bangladesh, Burma and Srilanka. *Zool.surv.India.Calcutta*.

Kamal, M. A. and Parvez, I. (2011). Molecular markes for the analysis of genetic biodiversity with special reference to Indian fish fauna. In: *Animal diversity, natural history and conservation.* Gupta, V.K. & Verma, A.K. (eds). Daya Pbl. Hum. New-Delhi. Vo. I: 249-274.

Kar, D, (2003) Fishes of Barak Drainage, Mizoram & Tripura. PP 203-211. Kumar, A. and Singh, L.K. (ed). In:Environment pollution & management. APH publ.corp. New Delhi.

Kar, D, (2005). Fishes of Barak Drainage, Mizoram & Tripura With a Note on Conservation. *J. freshw.biol.16.*

Khan, H.A. and Kamal, M. Y. (1979). On a collection of fish from river Kosi (Bihar). J. Bombay. Nat. Hist. Soc. 76 (3):530-534.

Nath, S. (1986). A checklist of fishes of Jammu & Kashmir state (India) with Remarks on the Ichthyography of the state. J.Zool.Soc.India. 38(1-2):83-98.

Sharma, B.D., (1978). Some new records of Fish from Jammu Province (J&K state). Livestock. Adv.3(11):31-32.

Silas, E.G. (1960). Fishes from the Kashmir Valley. J. Bombay Nat.Hist.Soc.57(1):66-77.

WCMC. (1998). Freshwater Biodiversity. A Preliminary global Assessment, A document Prepared for the 4th meeting of the Conference of the Practices to the Conservation of the Biological Diversity. World Conservation Management Centre.

❏❏❏

8

ORNAMENTAL AND FOOD FISH DIVERSITY OF TUNGA AND BHADRA RIVERS, WESTERN GHATS, KARNATAKA, INDIA

Shahnawaz Ahmad and M. Venkateshwarlu

ABSTRACT

The ornamental and food fishes are gaining paramount importance because of their economic and aesthetic values respectively. The study presents the documentation of ornamental and food fishes of Tunga and Bhadra Rivers, Western Ghats, Karnataka, India. Among 77 fish species recorded, 43 were categorized as food fishes, 29 as ornamental and 5 were not assessed due to lack of data. Hence, the present study reveals that this reverine system supports a rich fish fauna and recommends the need for effective and sustainable conservation measures to save such ornamental and food fishes from over-exploitation.

INTRODUCTION

Fish is a valuable source of food for millions of people as they serve as an excellent source of proteins for an increasing population in developing countries (Belsare, 1984). The extensive use of fish is more sustainable of livelihood of the rural poor (Belsare, 1986), because it serves as security of food for health and earned income. Besides, having important ornamental and food values, fishes are used as indicators of pollution in the aquatic eco-systems.

Attractive coloration, small size and suitability for keeping in captivity have attracted attention of the people all over the world towards ornamental fish trade. The percentage for ornamental fish trade has gone up to 85% for fresh water and rest is for marine fishes, invertebrates as well as fishes of cold and brakish waters (Dayal and Kapoor, 2000). The domestic market in India for ornamental and food fishes is growing well and the major leading centers include *Bombay, Madras, Cochin, Calcutta* and *Madurai*. India earns good money from the ornamental fish trade which accounts for more than Rs. 10 crores annually (Day, 1993). The Western Ghats (WG) located along the southwest coastline of the Indian subcontinent, is one of the biodiversity 'hotspots' of the world (Myers *et al.*, 2000). More recently, Shahnawaz *et al.* (2010): Shahnawaz *et al.* (2011) have presented a detailed account on the fish diversity, endemism, threat status and conservation measures of fishes in the Tunga and Bhadra rivers of Western Ghats, India. The aim of the study is to present the checklist of the ornamental and food fishes of the Tunga and the Bhadra rivers of the Western Ghats.

MATERIAL AND METHODS

During the present study, fourteen (14) study sites were selected from the Tunga and the Bhadra rivers, Karnataka, Western Ghats (Fig. 1). The selected sites were sampled for fishes during the period for June 2006 to June 2008. The fishes were sampled using various types for nets viz. gill nets of various mesh sizes ranging from 6-15 mm, drag nets, scoop nets and cast nets. Fishes were examined, counted and then released back into the system after taking few specimens (5-10) which were preserved in buffered formalin (10%) and transported carefully to the laboratory for further analysis. Fishes were identified based on keys for Fishes of the Indian Subcontinent (Jayaram 1999; Talwar and Jhingran 1991; Dutta Munshi and Srivastava, 2006).

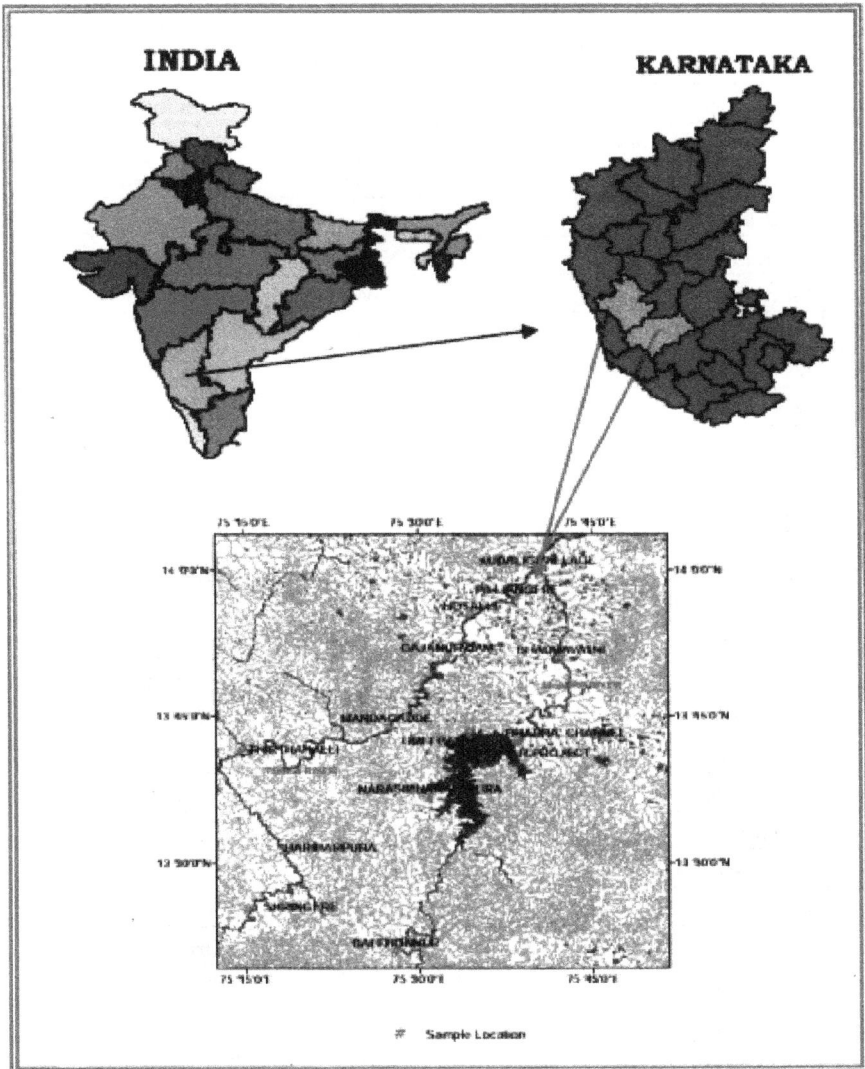

FIG. 1. Location of the study area.

RESULTS AND DISCUSSION

During the present study, a total of 77 fish species represented by 5 orders, 18 families and 44 genera were recorded. Among 77 fish species, 43 were categorized as food fishes, 29 as ornamental and rest 5 were not assessed due to lack of scientific data (Table 1)

Table 1. Endemic status and ornamental/ food fishes of the Tunga and the Bhadra rivers

Sl. No.	Fish species	Endemic status	Ornamental /food fishes
1	Osteobrama neilli	EN-WG	Food fish
2	Osteobrama cotio peninsularis	EN-WG	Ornamental
3	Puntius chola	EN-IS	Ornamental
4	Puntius arulius	EN-WG	Ornamental
5	Puntius sophore	EN-IS	Ornamental
6	Puntius jerdoni	EN-WG	Ornamental
7	Puntius sahyadrensis	EN-WG	Ornamental
8	Puntius filamentosus	EN-WG	Ornamental
9	Puntius ticto	EN-IS	Ornamental
10	Puntius amphibious	EN-I	Ornamental
11	Puntius sarana sabnastus	EN-WG	Food fish
12	Puntius conchonius	EN-IS	Ornamental
13	Labeo calbasu	EN-IS	Food fish
14	Labeo potail	EN-WG	Food fish
15	Labeo rohita	EN-IS	Food fish
16	Labeo porcellus	EN-WG	Food fish
17	Labeo angra	EN-I	Food fish
18	Labeo spp.	NA	NA
19	Cirrhinus fulungee	EN-WG	Food fish
20	Cirrhinus mrigala	EN-I	Food fish
21	Cirrhinus reba	EN-IS	Food fish
22	Rohtee ogillii	EN-WG	Ornamental
23	Danio malabaricus	EN-WG	Ornamental
24	Danio aequipinnatus	EN-IS	Ornamental
25	Danio rerio	EN-I	Ornamental
26	Tor mussullah	EN-WG	Food fish
27	Tor khudree	EN-WG	Food fish
28	Barilius gatensis	EN-WG	Ornamental
29	Barilius canarensis	EN-WG	Ornamental
30	Barilius bendelisis	EN-I	Ornamental
31	Garra bicornuta	EN-WG	Food fish
32	Garra mullya	EN-I	Ornamental

Contd...

Table 1. Contd...

33	*Hypselobarbus kolus*	EN-WG	Food fish
34	*Hypselobarbus thomassi*	EN-WG	Food fish
35	*Hypselobarbus lithopidos*	EN-I	Food fish
36	*Osteochilichthys thomassi*	EN-WG	Food fish
37	*Osteochilichthys nashii*	EN-WG	Food fish
38	*Salmostoma boopis*	EN-WG	Ornamental
39	*Salmostoma sardinella*	EN-I	Food fish
40	*Amblypharyngodon mola*	EN-I	Ornamental
41	*Rasbora daniconius*	EN-I	Ornamental
42	*Psilorhynchus tenura*	NA	NA
43	*Cyprinus carpio cummunis*	EX	Food fish
44	*Catla catla*	EN-IS	Food fish
45	*Balitora mysorensis*	EN-WG	Ornamental
46	*Schistura semiarmatus*	EN-WG	NA
47	*Nemachili chthys rueppelli*	EN-I	Ornamental
48	*Lepidocephalus thermalis*	EN-I	NA
49	*Mystus cavasius*	EN-IS	Food fish
50	*Mystus armatus*	EN-WG	Food fish
51	*Sperata aor*	EN-IS	Food fish
52	*Sperata seenghala*	EN-IS	Food fish
53	*Rita gogra*	EN-IS	Food fish
54	*Rita pavimentatus*	EN-WG	Food fish
55	*Mystus krishnensis*	EN-WG	Food fish
56	*Mystus malabaricus*	EN-WG	Food fish
57	*Ompok pabo*	EN-I	Food fish
58	*Ompok bimaculatus*	EN-IS	Food fish
59	*Wallago attu*	EN-IS	Food fish
60	*Clarias batrachus*	EN-I	Food fish
61	*Heteropneustes fossilis*	EN-IS	Food fish
62	*Xenentodon cancilla*	EN-IS	Food fish
63	*Parambassis thomassi*	EN-WG	Ornamental
64	*Chanda nama*	EN-IS	Ornamental
65	*Etroplus maculatus*	EN-WG	Ornamental
66	*Oreochromis mossambicus*	EX	Food fish
67	*Glossogobius giuris*	EN-IS	Food fish
68	*Nangra itchkeea*	EN-WG	Food fish
69	*Glyptothorax lonah*	EN-WG	Ornamental
70	*Proeutropiichthys taakree*	EN-WG	Food fish

Contd...

Table 1. Contd...

71	*Neotropius khavalchor*	EN-I	NA
72	*Botia striata*	EN-WG	Ornamental
73	*Channa punctatus*	EN-I	Food fish
74	*Channa marulius*	EN-I	Food fish
75	*Aplocheilus lineatus*	EN-I	Ornamental
76	*Mastacembelus armatus*	EN-IS	Food fish
77	*Notopterus notopterus*	EN-IS	Food fish

EN-WG: Endemic to Western Ghats; EN-IS: Endemic to Indian Subcontinent:

EN-I: Endemic to India; EX: Exotic; NA: Not assessed

and Fig. 2. The fish fauna of these two rivers consists of craps, catfishes, air-breathing and other fishes. Among the recorded fishes, *Catla catla, Cyprinus carpio, Oreochromis mosambicus, Sperata seenghala, Sperataaor, Channa marulius, Channa punctatus, Labeo rohita, Cirrhinus mrigala and Notopterus notopterus* are economically important and are an important component of national income. The study also shows that these rivers are rich in ornamental fish diversity. The ornamental fishes include *Puntius chola, Puntius arulius, Puntius sophore, Puntius sahyadrensis, Puntius filamentosus, Puntius ticto, Puntius amphibius, Puntius conchonius, Rohtee ogilbii, Osteochilichthys thomassii Osteochilichthys nashii, Basrilius gatensis, Barilius canarensis, Barilius bendelisis, Botia striata, Danio rerio.* Both the ornamental and food fishes were represented mainly by the family *Cyprinidae followed* by *Balitoridae* (Loaches) and *Bagridae- Catfishes* (Bhat, 2003; Shahnawaz *et al.*, 2010). Among the ornamental fishes, *Puntius conchonius, P. filamentosus, Barilius canarensis, Barilius bendelisis, Botia striata* and *Danio rerio* are the most commonly marketed aquarium fish species. Due to over-exploitation, habitat loss and aquarium trade makes these fishes vulnerable and therefore, the present study suggests strong and potent check and conservation measures of these rivers.

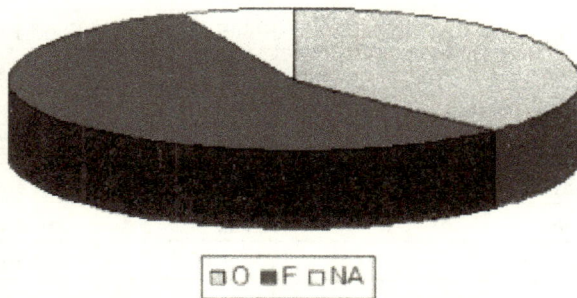

FIG. 2.   Ornamental/food fishes of the Tunga and the Bhadra rivers, Western Ghats.

REFERENCES

Belsare, D. K. (1984). Fish production in tropical waters as protein and food sources in developing countries. *In UNEP Bulletin on Ecological Approaches to Resources* (Ed. E. Seidel), 12:81-162.

Belsare, D. K. (1986). Tropical Fish Fanning. Karad (Maharashtra) Finlayson, (M.Presidents Report). *Wetlands*, 12:2.

Bhatt, A. (2003). Diversity and composition fo freshwater fishes in river systems of Central Western Ghats, India. Env. Biol. Fish, 68:25-38.

Datta Munshi, J.S. and M. P. Srivastava. (2006). Natural history of fishes and systematic of freshwater fishes of India. Narendra Publishing House, Delhi, p. 394.

Dayal, R and Kapoor, D. (2000). Survey of Existing Database of Endangered Species of Peninsular India. 192-193, NBFGR, NATP Publication, p. 278.

Dey, V.K., (1993). Ornamental fishes-Handbook on aquafarming. MPEDA Publication, Marine products, Export Development Authority, Kochi-682036, India, p. 76.

Jayaram, K.C., (1999). The freshwater fishes of the Indian region. Narendra Publishing House, Delhi-6.

Myers, N., R.Mittermeiner, G. C. Mittermeier, G.A.B. Dafonseca and J. Kent. (2000). Biodiversity hotspots for conversation priorities. Nature, 403:853-858.

Shahnawaz A., Venkateshwarlu, M., D.S. Somashekar and K. Santosh. Fish diversity with relation to water quality of Bhadra River of Western Ghats (INDIA). Environ Monit Assess, (2010), 161:83-91. DOI 10.1007/s10661-008-0729-0.

Talwar, P.K. and A. Jhingran. (1991). Inland fishes of India and adjacent countries. Oxford and IBH Publishing Co. Pvt. Ltd., New Delhi, 2(19): 1158.

❑❑❑

9

ON THE DIEL FLUCTUATIONS OF SOME PHYSICO-CHEMICAL PARAMETERS AND TROPHIC STATUS OF LAKE MANSAR

Anil Khajuria

ABSTRACT

Variations in physico-chemical parameters in lake Mansar studied during different seasons of the year revealed marked diel periodicity in respect of some of the factors analyzed. Available data also suggest that lake under investigation is productive and moving faster towards eutrophication.

INTRODUCTION

Mansar lake is situated 65 km east of Jammu city (32.42′N – 75.23′E) at an altitude of 710 meters above mean sea level. Besides, being a habitat of wintering and transit water fowls, the lake has abundance of biotic communities viz., planktons, algae, benthos, fishes and macrophytes. These biotic communities are profoundly influenced by the seasonal as well as diel variations in physico-chemical parameters of the water body. Present communication is an attempt to analyze the diel fluctuations in some physico-chemical parameters and their bearing on the trophic status of the lake.

MATERIALS AND MEATHOD

Analysis of water samples were made at 4 hour interval for a period of 24 hrs in the months of January, May, July and November. The temperature was measured by mercury filled Celsius thermometer while pH was recorded with portable pH meter. Chemical analyses for dissolved oxygen, free carbon dioxide, carbonates, bicarbonates, calcium, magnesium and chloride were made adopting standard methods (APHA, 1975 and Tridev *et al.*, 1987).

OBSERVATIONS AND DISCUSSION

Detailed observations on mean diel variations on some physico-chemical parameters in the months of January, May, July and November are depicted in Tables 1-4. Both the air and water temperature showed a definite ascending trend during daytime and descending trend during night hours. Diel changes in dissolved oxygen contents revealed the lowest ebb during night (2000-2400) and the highest during the noon and afternoon in 24 hrs cycle (Tables 1-4). More or less similar observations have been reported by Singh (1989) and Ahmad and Singh (1990) in the water bodies studied by them. While afternoon rise in dissolved oxygen content is attributed to active photosynthesis (Goldman 1968, Sumitra Vijayragvan 1971 and Dutta and Malhotra 1987), nocturnal fall reflects its consumption in respiratory activity of biota (Khajuria 1993).

pH has been observed to rise in its value from dawn onwards and fall from dusk which again seems to be the consequence of photosynthetic activity in the lake. Similar rise and fall in pH value has been recorded by Singh (1989) and Ahmad and Singh (1990). Free carbon dioxide attains its diel maxima during night but remains completely absent during day. Such diel fluctuations in free carbon dioxide as observed presently, is suggestive of its inverse relationship with that of dissolved oxygen which is already on record (Jhingran 1982, Malhotra *et al.*, 1984, 1987, Singh 1989 and Kumar 1990).

The diurnal cycle of carbonate and bicarbonate suggest alteration of carbonate and bicarbonate. Carbonate increase as a consequence of increased photosynthesis during sun shine hours when bicarbonate is used as a source of carbon releasing carbonate. While carbon dioxide produced through respiratory activity of plants and animals during night coverts carbonate back to bicarbonate and thus fall in concentration of carbonate consequently results in rise of bicarbonate contents (Tables 1-4).

Table 1 : Mean diel variations of physico–chemical parameters during the month of January.

Parameters	HOURS					
	0400	0800	1200	1600	2000	2400
Air Temp. (°C)	6.25	9.37	20.57	17.00	9.43	8.62
Water Temp. (°C)	10.87	11.68	17.37	15.25	13.17	11.43
DO (mg/l)	5.63	5.47	7.82	7.78	7.29	4.93
pH	7.64	8.02	8.02	8.52	7.44	7.11
FCO_2 (mg/l)	5.13	5.13	2.23	A	4.69	4.01
CCO_3- (mg/l)	A	A	4.51	10.84	A	A
HCO_3- (mg/l)	154.33	147.44	127.87	82.67	150.50	152.96
Cl^- (mg/l)	1.09	1.19	1.11	1.19	1.21	1.00
Ca^{++} (mg/l)	41.38	38.08	35.98	39.14	40.47	47.83
Mg^{++} (mg/l)	9.42	9.26	7.36	7.17	9.71	9.65

Table 2 : Mean diel variations of physico – chemical parameters during the month of May.

Parameters	HOURS					
	0400	0800	1200	1600	2000	2400
Air Temp. (°C)	29.50	35.25	34.87	30.72	25.37	20.17
Water Temp. (°C)	27.12	27.37	30.50	27.65	25.00	24.25
DO (mg/l)	8.56	9.40	12.17	12.00	5.94	4.60

Contd...

Table 2 : Contd...

pH	8.41	8.43	8.67	8.77	8.30	8.30
FCO_2 (mg/l)	0.10	0.03	0.05	A	0.05	0.05
CCO_3^- (mg/l)	4.07	4.10	7.42	6.55	1.67	4.75
HCO_3^- (mg/l)	135.75	109.82	117.77	128.87	138.25	135.17
Cl^- (mg/l)	0.92	0.93	1.14	0.96	0.88	0.85
Ca^{++} (mg/l)	25.50	21.52	24.67	23.75	24.77	27.95
Mg^{++} (mg/l)	6.11	5.64	5.78	6.76	7.05	7.27

Table 3 : Mean diel variations of physico-chemical parameters during the month of July.

Parameters	HOURS					
	0400	0800	1200	1600	2000	2400
Air Temp. (°C)	29.75	29.25	29.75	28.62	28.12	27.87
Water Temp. (°C)	26.80	27.37	27.00	26.00	25.00	26.37
DO (mg/l)	7.73	8.78	8.50	8.65	7.02	5.97
pH	8.31	8.37	8.35	8.36	8.15	8.07
FCO_2 (mg/l)	A	A	A	A	1.41	1.51
CCO_3^- (mg/l)	10.8	15.37	10.15	11.12	8.55	8.77
HCO_3^- (mg/l)	127.35	106.87	120.67	129.70	112.85	124.49
Cl^- (mg/l)	1.44	1.05	0.92	1.04	1.03	1.18
Ca^{++} (mg/l)	19.32	21.00	22.35	21.45	25.07	26.40
Mg^{++} (mg/l)	6.92	6.50	5.94	7.85	9.72	7.82

Table 4 : Mean diel variations of physico-chemical parameters during the month of November.

Parameters	HOURS					
	0400	0800	1200	1600	2000	2400
Air Temp. (°C)	12.82	14.87	17.17	22.10	14.62	13.25
Water Temp. (°C)	17.55	17.52	19.37	20.40	40.87	17.00
DO (mg/l)	6.49	7.54	8.17	7.38	7.32	7.32
pH	8.09	8.36	8.36	8.36	8.08	8.07
FCO_2 (mg/l)	6.74	1.33	1.11	0.44	7.13	4.22
CCO_3^- (mg/l)	0.96	2.76	3.03	4.09	A	0.96
HCO_3^- (mg/l)	112.75	108.34	110.08	99.52	126.48	115.70
Cl^- (mg/l)	1.03	1.08	1.92	1.72	1.74	1.84
Ca^{++} (mg/l)	27.68	27.68	28.08	27.05	30.32	30.87
Mg^{++} (mg/l)	5.88	5.71	6.51	6.37	8.99	7.84

Calcium and magnesium record nocturnal maxima and diurnal minima (Tables 1-4) which is very much similar to the pattern reported by Dutta and Malhotra (1987) and Kumar (1990) in their respective water bodies. Chloride contents do not experience any significant diel changes in their values except for little diurnal rise.

From overall pattern of diel fluctuations in different physico-chemical parameters authors conclude that (i) the rate of two mutually opposed processes viz., photosynthetic production of oxygen and its consumption for respiratory activities at any given time during 24 hrs cycle regulate the diel cycle of gasses and other associated chemical parameters in Mansar, (ii) there are wide fluctuation in physico-chemical parameters during 24 hrs cycle in lake Mansar. Relative productivity of a water body can be evaluated from the magnitude of fluctuations (Singh 1989, chaudhary et al. 1991 and Khajuria 1993). Wider the variations more productive is the water body. On the basis of data on the diel fluctuations in physico-chemical parameters, it may be deduced that the water body is productive rather moving fast towards eutrophication.

FIG. 1.   Lake Mansar Jammu an important tourist destination.

ACKNOWLEDGEMENT

This investigation was funded by Department of Environment, Govt of India. The author is indebted to Aayush Khajuria for Typing the manuscript and to Prof. B.L. Kaul for the photograph of the lake Mansar.

REFERENCES

Ahmad, SH and A. K. Singh (1990). Diurnal fluctuation of limno-chemical parameters in a freshwater pond of Dholi (Bihar), India. J. Hydrobiol. 6 (1) : 33-37.

A.P.H. A. (1975). Standard methods for the examinations of water and waste water, 12 th Amer. Publ. Hlth. Assoc. Inc., New York.

Choudhary, S. K., R. B. Singh, N. Manta, S. Choubey (1991). Diurnal profile of some physico-chemical and biological parameters in certain perennial pond and river Ganga at Bhagalpur (Bihar). J. Freshwater Biol. 4 (1):45-51.

Dutta, S.P.S. (1978). Limnology of Gadigarh stream with Special reference to consumer Inhabiting the stream. Ph. D. thesis, University of Jammu, Jammu.

Dutta, S. P. S. an d Y. R. Malhotra (1987). Diel variations in some hydrobiological parameters of Gadigarh stream, Jammu. J. Hydrobiol., 3(6) : 43-50.

Goldman, C.R. (1968). Aquatic primary production. Am. Zoologists, 8 :31-42.

Jhingran, V. G. (1982). Fish and Fisheries of India. Hindustan Publishing Corporation, India.

Khajuria, A. (1993). Studies on benthos and nektons of lake mansard. Ph.D. thesis, University of Jammu, Jammu.

Kumar, N. (1990). Altitude related limnological variations in some fish ponds of Jammu province. Ph. D. thesis, University of Jammu, Jammu.

Malhotra, Y.R., S.P.S. Dutta and S. N. Suri. (1984). Diurnal variations in physico-chemical parameters of a fish pond at Jammu. Ind. J. Ecology. 11 (2) : 339-341.

Sehgal, H.S. (1980). Limnology of lake Surinsar, Jammu with special reference to zooplanktons and fishery prospects. Ph. D. thesis, University of Jammu, Jammu.

Singh, D.N. (1989). Diel variations of certain physico-chemical parameters in Mcpherson lake, Allahabad. Proc., Nat. Acad. Sci., India 59 (3) IV: 363-367.

Singh, S. D. (1977). Some observations of diurnal vertical fluctuations in a pond having permanent bloom of *Microcystis flosaque*. J. Inland Fish. Soc., India 9 : 125-135.

Sumitra, Vijayraghavan (1971). Studies on diurnal variations in physico-chemical and biological parameters in Teppa Kulam Tank. Proc. Indian Acad. Sci., 74 B (20) : 63-73.

Trivedy, R.K., R.K. Goel, and C.L. Trisal (1987). Practical method in Ecology and Environmental Science.

❑❑❑

10

ZOOPLANKTON DIVERSITY OF SADER MOUJ RESERVOIR, BUDGAM DISTRICT, JAMMU AND KASHMIR

Ishrat Bashir, Md. Yousuf, Shahnawaz Ahmad and Shahshikanth Majagi

ABSTRACT

The studies on diversity and abundance of Zooplankton were carried out in Yusmarg Sader Mouj Reservoir. Samples were collected from four sites of the study area. Two samplings were done representing two seasons of the year. About 11 species were recorded representing six families. Of them Cladocera represented maximum number of species. Maximum number of individuals were recorded in the summer season compared to winter season. Shannon, Simpson and other diversity indices are presented.

Key words: *Sader Mouj Reservoir, Zooplankton, Rotifera, Cladocera, Copepoda Shannon diversity index.*

INTRODUCTION

Water is a resource that is essential for the survival of all species. Water supply impacts human and animal habitats, economic prosperity, food supply, human migration and domestic and international relations. Water resources are sources of water that are useful or potentially useful to humans. Uses of water include agricultural, industrial, household, recreational and environmental activities. Virtually all of these human uses require fresh water.

Aquatic habitats are inhabited by a variety of living organisms. Of these planktonic communities, because of their short life cycles, respond quickly to environmental changes and hence their standing crop and species composition indicate the quality of water in which they live (APHA, 1988).

Reservoirs are considered favorable environments to the development of plankton communities which may establish diversed assemblages in relatively short periods of time after impoundment (Rocha et.al. 1999). Several factors usually contribute to the establishment of plankton communities in a reservoir, among which are water quality, presence of nutrients physico-chemical factors of water, hydrological characteristics of the reservoirs and reservoir ageing. Phytoplanktons are usually at the base aquatic food web and are the most important factor for the production of organic matter in the aquatic ecosystem. Most reservoirs require significant amount of plankton to have productive and sustainable fisheries. The interplay of physical, chemical and biological properties of water most often lead to the production of phytoplankton while their assemblages are structured by these factors. Any perturbations in the factors may effect their assemblages which could have a significant impact on water quality and fisheries of reservoirs. The zooplankton assemblages often influence flow through classical food chain, nutrient cycling and community population dynamics within a reservoir system. This ecological niche has made them key in top down grazing effect (trophic cascade) on the bottom up forces which plays pivotal roles in bio-manipulation for lake restoration purpose (Carpenter and Kitchell, 1993).

Forbes (1887) described the physico-chemical and biological composition of standing water bodies and designated standing water bodies as microcosms. In the same year another British Ecologist Victor Henson (1887) coined a term 'Plankton' and used the word to describe all microscopic free floating organism of aquatic system.

Zooplanktons are at the base of the food chain, feeding on microscopic plants and being fed upon by aquatic insects and planktivorous fishes. Zooplanktons have potential value as assessors of trophic conditions. They respond more rapidly to environmental changes thus proving to be good indicators of water quality conditions. Functionally, the zooplankton includes detritivores, herbivores, carnivores, all of which secrete dissolved and particulate matter (both organic and inorganic) that can serve as nutrient for saprovores and phytoplankton.

According to Sprules and Munawar (1991). Lindegaard (1994) Plankton and zoobenthos play an important role in lake ecosystems as a main determinant of hydro biological production and community structure. Jacqueline and Kenneth (1994) opined that the interaction between zooplankton and phytoplankton forms an important basis of food chain in natural lakes according to Alois Herzing (1987) and Zaret (1980) the former also serves as an essential source of macro invertebrates. The plankton study is a very useful tool for the assessment of biotic potential and contributes to overall estimation of basic nature and general economic potential of water body Pawar et al (2006).

The most notable studies of the ecology of plankton in high altitude Andean Lakes occurred during the late seventies. They included surveys (Widmer *et.al.* 1975), Preliminary studies of the relationship between zooplankton and environment (Colvinaux and Steinitz 1980) and preliminary assessment of trophic relationship affecting the diversity of zooplankton (Hurlbert et.al. 1986).

The present investigation was undertaken to study Zooplankton in Sarder Mouj Reservoir, a minor water reservoir in Yusmarg region of Jammu and Kashmir, state of India. The study aimed at:

- Conducting a preliminary survey of Zooplankton density and taxonomic composition in Yusmarg water reservoir,
- Assessment of relative abundance and seasonal dynamics of zooplankton over a relatively short time-frame, and
- Providing baseline data on the zooplankton community of Yusmarg water reservoir.

STUDY AREA AND STUDY SITES

Yusmarg is small meadow set in the heart of mountains to the south-west of Srinagar and is approximately 47 km from Srinagar. It lies in the Budgam district of Jammu and Kashmir at an altitude of approximately 2,712 m above mean sea level within the geographic coordinates of 33°49'42" N latitude and 74°39'59" E longitudes.

The area of Yusmarg (Fig. 1) enjoys a sub-temperate climate with the basic seasons of summer and winter. Precipitation in Yusmarg is normally in the form of mild snowfall during the winters. Summers are mild and winters in Yusmarg are very cold. The maximum temperature ranges around 30°C during the summer months. During winter months Yusmarg experiences a maximum temperature of 15°C to 8°C and a minimum temperature of around – 2°C.

Table 1: Representing features of Sader Mouj water reservoir.

S.No.	Features of water reservoir
1	Approximately 34 years old.
2	Depth : 19.81 m.
3	Area: 27.18 acres.
4	Bottom: Silty (inlet-sandy).
5	Fish type: Mirror carp.
6	Inlet supply from: Khansha Maansha Canal (approx. 30 kms. From the reservoir, real source-in Pir Panjal Mountains.
7	Built to address drinking water supplies and irrigation
8	Irrigation to : Pakharpora, Dalwan, Foutlipora, Zenpanchal, kadipora (District – Pulwama), Aaglaar (District Pulwama) areas.
9	Reservoir filtration plant at: Monasarbandh

STUDY SITES

For studying the zooplankton community the reservoir was sampled at four sites.

Site I – Outlet

It lies near the outlet of the reservoir at an altitude of 2,363 m above mean sea level within the geographic coordinates of 33°82'40.6 N latitude and 7467'37.3 E longitude.

Site II – Forest side

This site is near the edge of forest that is dominated by tree species of *Pinus wallichiana* and *Abies pindrow*. It lies at an altitude of 2,360 m above mean sea level within the geographic coordinates of 33° 49' 32.3 N latitude and 74°40'04.1 E longitude.

Site III- Road side

This site is located just near the roadside. It lies at an altitude of 2,373 m above mean sea level within the geographic coordinates of 33°49' 32.4 N latitude and 74°40'10.3 E longitude.

Site IV- Inlet

It lies near the inlet of the reservoir towards the periphery. It lies at an altitude of 2,364 m above mean sea level within the geographic coordinates of 33°49'93.89" N latitude and 74°39'57.8 E longitude.

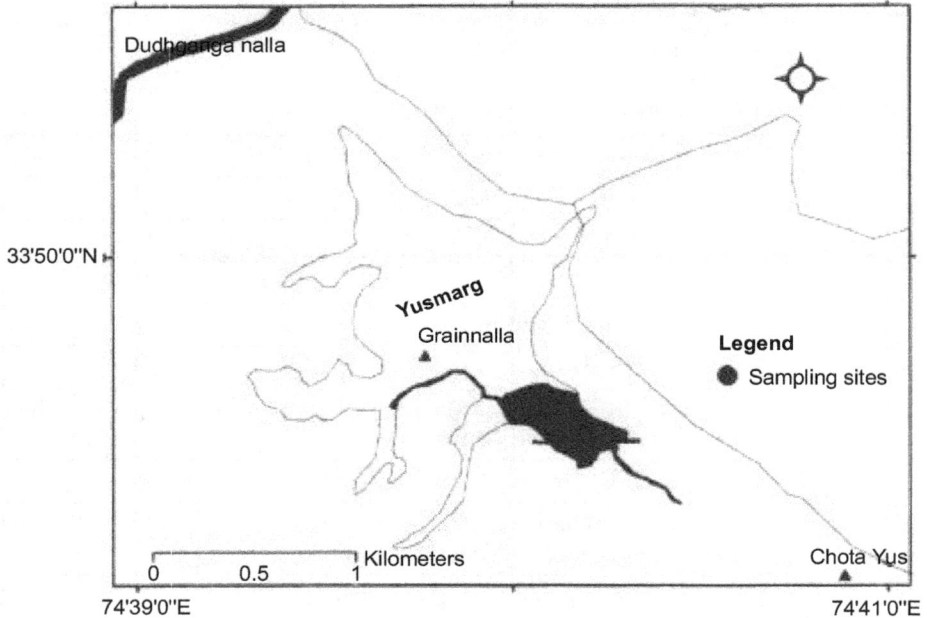

FIG. 1. Setellite image of Yusmarg area with 4 study sites

MATERIAL AND METHODS

Qualitative Study of Zooplankton

For the qualitative enumeration of zooplankton, samples were collected with the help of plankton net of bolting silk (mesh size no. 25) by making horizontal as well as vertical hauls. The contents, collected in the plankton tube attached to the lower end of the net, were transferred to separate polyethylene marked tubes and preserved by fixing in 4% formalin. The plankton samples were reduced in volume by centrifuging for about 10 minutes. Identification of organisms was done under microscope with the help of standard taxonomic works on the group done by Edmondson (1959); APHA (1998).

Quantitative Study of Zooplankton

For quantitative study of zooplankton, 10 liters of water collected were sieved through plankton net. The contents were preserved in 4% formalin. The plankton samples were concentrated and shaken thoroughly. While shaking one ml of the sample was placed in the Sedgwick rafter cell and studied under microscope. The whole cell was scanned for different zooplankton. The counts were made in triplicate and average of three counts was used to calculate the number of various genera and total zooplankton per cubic meter of the water by the formula.

$$n = ac/I \times 1000$$

where;

n = no. of individuals per cubic meter of the water,

a = average no. of individuals of the three readings,

c = volume of the concentrated sample (in ml),

I = volume of the original water sample sieved (usually 10 liters).

Besides, Diversity index (H) of zooplankton community was calculated according to Shannon-Weaver Index:

$$(H) = -\Sigma\left[(n_i/N)\, ln\,(n_i/N)\right]$$

where;

H = Shannon – Weaver Index

n_i = no. of species

N = total no. of species

The higher value of H, the greater the species diversity of the community.

The similarity between the communities of zooplankton in the 4 sites of the reservoir was determined using Sorenson's index:

$$S = 2c/a + b \times 100$$

where;

c = no. of species common to both sites

a = no. of species in one site

b = no. of species in another site

S = similarity index between two Sites

RESULTS

The General features of Sader Mouj reservoir are depicted in Table No. 1. Zooplankton diversity of Sarder Mouj reservoir recorded from May 2010 to December 2010 at four different sites and four species of Rotifera, five species of Cladocera and two species of copepod during two seasons i.e. summer season and winter season of the study period.

The results depicted in the Tables 2 to 5 the rotifer species were dominant 715 in site 1 and lowest 134 in site 4. Similarly Cladocera and Copepod recorded maximum 1452 and 704 in site 2 and minimum 103 and 52 in site 4 respectively.

Table 2: Zooplankton Density at various Sites of Yusmarg Water Reservoir from May, 2010 to December, 2010.

Taxa	Site 1					Site 2					Site 3					Site 4				
	May	June	Nov.	Dec.	Mean	May	June	Nov.	Dec.	Mean	May	June	Nov.	Dec.	Mean	May	June	Nov.	Dec.	Mean
Rotifera																				
Keratella sp.	40	40	16.66	16.66	112	80	80	10	11	181	80	86.66	11	15	192	16.66	16.66	3.33	0	35
Brachionus sp.	80	120	23.33	20	243	80	120	12	13	225	73.33	120	18	20	231	10	6.66	0	0	16
Platyias sp.	60	100	20	0	180	16.66	20	2	2	40	10	23.33	10	8	51	3.33	3.33	0	0	06
Anuraeopsis sp.	50	90	30	10	180	20	36.66	5	6	67	13.33	16.66	2	6	37	10	10	0	0	20
Total Rotifera	230	350	89.99	46.66	715	196.66	256.66	29	32	413	176.66	246.65	41	49	511	39.99	36.65	3.33	0	57
Cladocera																				
Bosmina sp.	100	63.33	30	26.66	229		60	28	24	228	160	96.66	38	25	319	16.66	23.33	3.33	0	42
Daphnia similis	183.33	240	100	30	553	163.33	206.66	108	96	573	163.33	206.66	80	76	525	10	23.33	6.66	0	39
Daphnia pulex	120	183.33	60	16.66	379	120	280	70	65	535	90	120	20	22	252	6.66	10	0	0	16
Graptoleberis sp.	20	56.66	20	0	96	20	50	15	2	87	10	20	2	2	34	3.33	3.33	0	0	6
Ceriodaphnia sp.	0	0	0	0	0	6.66	10	8	5	29	0	0	0	0	0	0	0	0	0	0
Total Cladocera	423.33	543.32	210	73.32	1257	429.99	606.66			1452	423.33	443.32				36.65	59.99	9.66	0	
Copepoda																				
Cyclops sp.	256.66	210	50	30	546	260	203.33	60	68	591	120	103.33	50	58	331	30	10	3.33	0	43
Copepodite	70	30	3.33	0	103	70	30	5	8	113	50	40	24	28	142	6.66	3.33	0	0	09
Total Copepoda	326.66	240	53.33	30	649	330	233.33	65	76	704	170	143.33	74	86	473	36.66	13.33	3.33	0	52
Total Zooplankton																				

*'-' = No Sampling   *'0' = Absent

Among the rotifer Brachionous species were dominant in all the three stations except site 4 where Keretelle species were dominant Among the Clodocera Daphnia *Siphilis* is dominant in 1,2 and 3. Whereas site 4 showed *Bosmania* species dominant. Similarly copepod shows Cyclops species in all the four sites, highest recorded in site 2 and lowest in site 4. Among all the rotifer *Brachionous* species recorded in site one and lowest *Keratella*, where as *Platyias* species in site 4 i.e. 06 in site 4. Similary in cladocera *Daphnia* dominance in all four stations, whereas Cariodaphnia is less dominant among all Cladocera in all stations. In station one and four they did not record. All zooplankton recorded highest in summer and lowest in winter in all stations. Table 3 depicts the relative density of various groups-cladocera is dominating group among the three groups. Table 4 shows the list of families of different zooplankton groups. Brchionidae and Daphniidae represent maximum species four and three respectively. Percentage contribution of different families of zooplankton of Sader Mouj reservoir is shown in Table 5.

Table 3: Relative Density of various zooplankton groups at different study sites.

Group/Site	1	2	3	4
Rotifera	27.38	22.07	26.40	33.33
Cladocera	47.77	50.49	54.05	44.44
Copepoda	24.85	27.44	19.55	22.22

Table 4: List of Zooplankton Species recorded in Yusmarg Water Reservoir from May, 2010 to December, 2010.

Family	Species
Bosminidae	*Bosmina* sp.
Brachionidae	*Anuraeopsis* sp. *Brachionus* sp. *Keratella* sp. *Platyias* sp.
Chydoridae	*Graptoleberis* sp.
Cyclopoidae	*Cyclops* sp.
Daphniidae	*Ceriodaphnia* sp. *Daphnia pulex* *Daphnia similis*
Copepodite – a developmental stage of copepods.	

Table 5: Percentage Contribution of different families of Zooplankton in Yusmarg Water Reservoir.

Family/ Site	1	2	3	4
Bosminidae	8.04	8.77	16.01	18.06
Brachionidae	27.39	22.07	26.40	33.33

Contd...

Table 5 : Contd...

Chydoridae	3.69	3.40	1.87	2.77
Copepodite	20.89	22.56	13.93	18.06
Cyclopoidae	3.95	4.87	5.61	4.16
Daphniidae	35.67	38.31	36.18	23.61

Table 6 : Ritifera

Indices	Site1	Site2	Site3	Site4
Individuals	715	513	511	77
Dominance_D	0.2668	0.34	0.3607	0.3233
Shannon_H	1.352	1.194	1.147	1.234
Simpson_1-D	0.7332	0.66	0.6393	0.6767
Evennes_e^H/S	0.966	0.8249	0.787	0.8586

Table 7 : Cladocera

Indices	Site1	Site2	Site3	Site4
Individuals	1257	1452	1130	103
Dominance_D	0.3235	0.3201	0.3462	0.3372
Shannon_H	1.229	1.272	1.153	1.188
Simpson_1-D	0.6765	0.6799	0.6538	0.6628
Evennes_e^H/S	0.8548	0.7138	0.7921	0.8204
Margalef	0.4204	0.5494	0.4267	0.6473

DISCUCSSION

The seasonal patterns are different in the study area. Here we experience only two seasons *i.e.* summer and winter. Summer from February to June and Winter from July to January accordingly values are interpreted.

Analyses of large number of lakes has revealed that mean zooplankton species richness was constrained by variations in lake latitude and altitude i.e. high latitude and altitude lakes had always low richness (Dag etal 2006). It is difficult to distinguish among immigration, climate or other spatially structured constraints as causes of geographic patterns in richness. The fact that most zooplankton have wide geographical distribution and moderate to good colonization abilities (Shurian 2000; Jensen et al 2001). Combined with long period for colonization since the last period of glaciations suggests that relatively low diversity observed in western lakes (65% relative to eastern sites) arise from biological controls of diversity, rather than on the inability of species to reach potential habitats (Havel and Shurian 2004).

In studies carried out on the Amazon lake (Hardy 1980). The Monjolinha reservoir (Naueira and Matsumura-Tundisi 1996) two lake located in the Pantanal; Albuquerque (Espmitola et al 1996) and Sinha Mariana (Morini Lopes 1999) and two marginal lakes of the river Paran (Nunes etal 1996) a smaller number of species were recorded despite the greater number of samples taken and the larger size of the water body (Neves et al 2003).

In the present study the greatest organism abundance was recorded in summer compared to winter season. Although only a single study was carried out for each site in each season yet it is possible that his pattern is representative for the seasons. In lakes without periodic flooding the dry season brings greater stability due to organic matter production and decomposition as exemplified in other Pantanal systems (Pinto-Silva 1991, Lima 1996) so that organism abundance can be greatest in that season.

In relation to the Cladocera the family Daphnoidae was representative with the greatest number of species. The richness of Daphnoidae with the occurrence of two species was in concurrence with a similar work carried out by Wisniewski *et al.* (2000) who opined that this situation is to be expected in the Majority of Shallow environments.

Sampio and Lopez (2000) recorded numerical dominance of Copepods which they attributed to the fact that these organisms are strategic or opportunist with small size, short life cycle and wide tolerance to variety of environmental factors (Gren 1972b, Robertson and Hardy 1984).

In Rotifers different diversity indices are shown in Table 6. The Shannon and Simpson diversity index of rotifers 1.352 and 0.6393 is highest at site 1 and lowest at site 3 and 1.147 and 0.66 at site 3 and 2 respectively. Deminance index is maximum 0.3607 at site 3 and low 0.2668 at site 1.

In Cladocera different diversity indices are shown in Table 7. The Shannon and Simpson Diversity index is 1.272 and 0.6799. It is highest at site 2 and lowest 1.153 and 0.6538 at site 3.

In the present study the uniformity of the population was low within sites and seasons. Similar observation were made by Green (1993) who Opined that it is an indication of environmental stress. The low diversity and dominance of single rotifer species could have been related to the organic matter level of the system due to domestic sewage input through the stream. According to Margelef (1983) organically enriched environments have lower diversity with a few species dominating. The dominant rotifer species was Brachionous a speices associated with such environments (Sladecek 1983; Borosso et al 1997).

CONCLUSION

The present study concludes that zooplankton community fluctuates according to the habitat change and variations in the physico-chemical parameters of the environment. The study clearly demonstrates the richness of the cladocera component of the zooplankton in Sader Mouj reservoir. Daphniedae family contricuted maximum percentage. The study has made an attempt and it may help for further monitoring of zooplankton community and its uniqueness of Sader mouj reservoir in future.

REFERENCES

APHA, (1988) *Standard methods for the examination of water and waste water*. 20ᵗʰ Editio. American Public Health Assoc. Washington, D.C.

Barroso G. F, Dias Jr. C and Guntel A (1997) Preliminary assessment of the eutrophication potential from sewage effluents for four waste water treatment plants in Espiroto Santo State (Brazil) Ver. Limnol. 26; 666-670.

Carpenter S. R. and Kitchell J F. (1993) The trophic cascades in lakes. Cambridge University Press, Cambridge, p. 385.

Colinvaux P and Steinitz M. (1980) Species richness and area in Galapagos and Andean lakes: Equilibrium phytoplankton communities and a paradox of the zooplankton, p 697-712. In W. C. Kerfoot (Ed.) Evolution and ecology of zooplankton communities. Unversity press of New England Honnover, New Hampshire.

Dag O. Hassen, Bjorn A., Faafeng Val H Smith, Vegar Bakkestaren and Bjorn Walseng: (2006). Ecology. 87 (2) pp. 433-443.

Edmonson, W. T. (1959) Freshwater Biology. John Wiley and Sons Inc, New York, London.

Espindola E. L. G. Matsumara-Tundsi T. and Morena I, II (1996): Effects of the hydrological dynamics of the Mato-grossan Panantal on the community structure of the Zooplankton of Lake Albuquerque. Acta Limnol. Brasil., 8:37-57 (In Protugeese).

Forbes, S. A. (1887) The lake as a microcosm. *Bull Sci. Assoc.*, Peoria, Illinois, pp. 77-87. Reprinted in *Illinois Nat. Hist. Survey Bulletin* 15(9):537-550.

Green J. (1927b) Fresh water ecology in the Mato-grosso Central Brazil III. Association of Rotifera in Meander lakes of the Rio Suia Missu, J. Nat. Hist. 6;229-241.

Green J. (1993) Diversity and abundance in planktonic rotifers. Hydrobiologia. 255/256: 345-352.

Hardy E, R. (1980) Zooplankton composition in the Amzaonian lakes Acta. Amazonica 10;557-609. (In Portuguese).

Harvel J. H. And Shurin J. B. (2004) Mechanisms, effects and scales of dispersal in freshwater zooplankton Limnology and Oceanography 49. 1229-1238.

Herzig, A. (1987) The analysis of plankton rotifer population: a plea for long-term investigations. Hydrobiologia 147:163-180.

Hurlbert, S. H., Loayza W. and Morena T. (1986) Fish Flamingo plankton interaction in the Peruvian Andes Limnol. And Oceanogr. 31: 457-468.

Jensen T. C., Hessen D. O. and Fanfeng B. A. (2001) Biotic and abiotic preferences of cladoceran invader *Limnosida frontosa* Hydrobiologia. 442: 863-866.

Lima D. (1996). Community structure of the zooplankton and phytoplankton of Lake Recreio Bardo de malgago Panantal Mato-grosso. Master's Thesis, Federal University of Sao Carlos, Sao Carlos, p. 209 (In Portuguse).

Lindegaard, C. (1994). The role of zoobenthos in energy flow in two shallow lakes. Hydrobiologia 275/276:313-322.

Marglef R. (1983) Limnology, Ediciones omega S A., Barcelona p. 1010 (In Spanish).

Morino-Lopes, A. A. E. T. (1999). Limnological conditions and zooplankton composition of Lake Sinha-Mariana, Bardo de Malgago, Mato-grossan Panatanal. Masters thesis, Institute of Biosciesnce, Federal University of Mato Grosso, Culaba p. 98 (In Portugese).

Neves I F. Rocha O. Roche K F. and Pinto A. A. (2003). Zooplankton community structure of two marginal lakes of the river Cuiaba (Mato-Grosso, Brazil) With analysis of Rotifera and Cladocera diversity. Braz. J. Biol. 63(2) 329- 343.

Nogueria M G. Matsumura-Tundsi T. (1996). Limnology of a shallow artificial system (reservoir of Monjolinho, Sao Carlos, SP) Population dynamics of the plankton (Acta Limnol. Brasil., 149-168. (in Portuguese).

Nunes, M. A., Lansac-Toiia, F. A. Bonecker, C.C. Roberto, M. U. and Rodrigues, L, (1990). Composition and abundance of the zooplankton in two lakes of the Horto Florestal Dr. LoizTelexiera Mendes, maringa, Paranna. Acta Limnol. Brasil., 8; 207-219 (in Portuguese).

Pawar, S. K., Pulle, J. S., and Shendge, K. M. (2006) The study on phytoplankton of Pethwadaj dam Tq. Kandhar, Dist. Nanded, Masharashtra. J. aqua Boil 21 (1): 1-6.

Pinto-Silva V. (1991): Diurnal variation of the principle limnological parameters In the lakes of the Recrerio and Buitizal Mato-Grossan Panatal, Bordo de Melgago-MT. Doctoral thesis, Federal University of Sao Carlos, Sao Carlos 126p. (In Portuguese).

Roberston B. A. and Hardy E. R. (1984). Zooplankton of Amazonian lakes and rivers, pp 337-352. In II Sioli (Ed.) The Amazon Limnology and landscape, Ecology of Mighty tropical river and its basin. W. Junk Publ. Netherlands.

Rocha O. Matsumur-tundsi, Espindola T. E. L. G., Roche K F. and Reitzler A.C. (1999) Ecological theory applied the reservoir plankton. In Tundsi J G. & M. Straskraba (eds.) Theoritical reservoir ecology and its applications. International Institute of Ecology, Brazilian Academy of Sciences, Backhuys publishers, Leiden, Neitherland, p. 29-51.

Sampio, E. V. and Lopez C.M. (2000). Zooplankton Community composition and some limnological aspects of an Oxbow lake of the Paraopeba River, Sao Fransisco river basin, Minas Gerias, Brazil Braz. Arch. Bio. Technol.43:285-293.

Shrin J. B. (2000) Dispersal limitation, invasion resistance and the structure of pond zooplankton communities. Ecology 81: 3074-3086.

Sladecek V. (1993): Rotifers as indicatorsof water quality. Hydrobiologia. 100; 169-201.

Sprules, W. G., Munawar, M. (1991) Plankton community structure in Lake St Clair, 1984. Hydrobiologia 219, 229-237.

Widmer C., Kittel T. and Richordson P. J. (1975) A survey of the biological limnology of Lake Titicaca. Verhandlungen. Internationale Vereinigung fuer theoretische und angewandte Limnoligie 19: 1504-1510.

Wisniewski M. J., Rocha O., Rietzler A. C. Espindola E. L. G. (2000). Diversity of zooplankton in Oxbow lakes of the Mogi-Guocu river flood planes II Caldocera (Crustecea, Brachipoda) pp. 559-586. In J. E. dos Santos & J. S. R. Pires (Eds.). Integrated ecosystem. Ecological stations of Jatai Vol.2 RiMa Editiors Sao Carlos (In Portugeuese).

Zaret, T. M. (1980). The effect of prey motion on planktivore choice. In : kerfoot WC (ed.) Evolution and ecology of zooplankton communities. Univ. Press New England Hanover, NH.

❏❏❏

11

A REVIEW ON THE MARINE FISH PARASITES OF RIO GRANDE DO NORTE, NORTHEASTERN BRAZIL

S. Chellappa, E.T.S. Cavalcanti, E.F.S. Costa, G.S. Araujo,
J.T.A. Ximenes-lima and N.T. Chellappa

ABSTRACT

Parasites are an integral part of all ecosystems and they represent an important factor in global biodiversity. Research in fish parasitology has increased exponentially in recent years owing to the expansion of aquaculture industry and the need to consume healthy fish stocks. This review focuses on the investigation of fish parasites carried out in the coastal waters of Northeastern Brazil and is based on available literature. Research has been focused on the occurrence, sites of infestation and seasonal variation of crustacean and nematode fish parasites. Lists of parasites and fish hosts are provided. Various parasites have been recorded for marine coastal fishes of this region. Parasites of zoonotic importance and host sex specificity in relation to gonadal parasitism have been reported. Much effort is still to be made on the life cycle of these organisms, their invasion and strategies of co-existence.

Key words: *Fish parasites, marine fish hosts, coastal waters of Northeastern Brazil.*

1. Why study marine fish parasitology?

Marine fish parasitology is an important field of study, due to its close linkage to fisheries, mariculture, ecology and environmental monitoring, Parasites are an integral part of all ecosystems, representing a major factor in global biodiversity. Host-parasite checklists suggest that on an average, there are at least 3 to 4 metazoan parasites per marine fish species and this led to a conservative estimate of 20,250 to 43,200 marine metazoan fish parasites (Klimpel et al., 2001; Plam, 2007).

Fish can be infected with various parasites, which can be either generalists (infecting different host species) or host-specific (infect only one or a few closely related host species). Fish may obtain parasites from their food or are directly infected by free-living parasitic stages. Parasites either have complex life-cycles, involving up to 3 or more different host speices, or direct lfie-cycle involving a single host. The most common fish parasites include skin crawlers (copepods), tongue biters (isopods), ectoparasitic flukes (monogeneans), endoparasitic flukes (digeneans), round worms (nematodes belonging to the family Anisakidae), tapeworms (cestodes) and spiny headed worms (acanthocephalans), among others (Williams, 1996; Rohde, 2005).

Crustacean ectoparasites on marine fishes are diverse. They often possess attachment organs that are deeply embedded in the host tissue. Other species that move freely on the surface of the fish can rupture the protective skin, destroy the mucus cover and open wounds for subsequent bacterial infections. Crustaceans together with monogenean flatworms are therefore the most problematic fish parasites in marine fish culture. Many fish species are infected by cymothoids (Crustacea: Isopoda: Cymothoidae), which feed on the blood of the host. Many species of parasites settle in the buccal cavity of fish, other live in the gill chambers or on the body surface including th fins. Their life cycle involves only one host (Holoxenic cycle). Isopods are associated with many species of commercially important fishes around the world and cause significant economic losses to fisheries by killing, stunting, or damaging fishes. Isopods serve as an important food item for a variety of animals, and are commonly seen as parasites on teleost fishes in tropical and subtropical waters. Parasitic isopods resemble free-living isopods except for their hook-like legs. The stages normally found are the non-swimming, permanently attached mature females, often with a small male nearby (Brusca, 1981; Bunkley- Williams and Williams Jr., 1998).

Fish-parasitic isopods are alleged to indicate tropical fishes which are free of ciguatera (fish poisoning toxins). This is not proven. Approximately 4,000 species have been described, and more than 450 species are known to parasitize fishes. They vary form 0.5 to 440 mm in length, and the largest species occurs off Puerto Rico (Bunkley-Williams et al., 2006).

The sea lice are crustacean copepods within the order Siphonostomatoida and family Caligidae. There are 36 genera within this family which include approximately 42 Lepeophtheirus and 300 Caligus species. Sea lice are marine ectoparasites that feed on the mucus. Epidermal tissue, and blood of host marine fish (Boxshall and Montu, 1997).

Cymothoid fish parasites have a dorso-ventrally flattened body with a head, fused with the first thoracic segment (cephalothorax), thorax and abdomen, Eggs, larval forms and juveniles develop in a brood pouch of the female. Free-swimming juveniles develop into adults. Sexes are separate in some isopods, while others begin life as males and turn into females (protandrous hermaphrodites). Fish-parasitic isopods feed on blood from wounds. Since cymothoids penetrate the host's skin with their pereopods and mouthparts, and the tissue-inhabiting forms maintain a small opening to the outside, microbial diseases. Pathogenic microorganisms in the aquatic habitat pose problems to the economic important fishes due to their secondary invasion on the body of the fish (Ravichandran et al., 2010).

Parasites can effect fish reproduction either directly or indirectly, depending on the target organ. A female with a muscle parasite load of 340 pseudocysts per gram has only 10 per cent of the fecundity of non-infected fish (Alderstein and Dorn, 1998). The parasite extracts energy reserves from the host, which are thus not destined to reproductive effort and induces physiological, immunological or ethological changes in the host. This impair mating, gonad maturation or larval survival. Goand infection can lead to parasitic castration, depending on the exact location and intensity of infection (Sitja-Bobadilla, 2009).

2. What are the different systematic groups of marine fish hosts and parasites registered in Rio Grande do Norte, Northeastern Brazil?

A pioneering study conducted during 2001/2002 registered the occurrence of metazoan ectoparasites on the commercially important marine fishes from the coastal waters of Rio Grande do Norte, Northeastern Brazil. A total of 687 fish samples, comprising of 16 families and 29 species, were necropsied and the parasites encountered were counted, processed using adequate techniques and identified. There were 90 fishes parasitized (Table 1). Of the parasitized marine fish species, only seven fish species were captured in sufficient numbers for analysis: Pomadasys corvinaeformis, Opisthonema oglinum, Pellona harroweri, Menticirrhus americanus, Chloroscombrus chrysurus, Polydactylus virginicus and Mugil curema. Maximum number of ectoparasites occurred on Pomadasys corvinaeformis (30.68%) followed by Mugil curema (21.59%), Chloroscombrus chrysurus (11.36%), Scomberomorus brasilensis (9.09%) and Oligoplites saurus (4.54%). The parasites encountered on the other fish species amounted to 22.74%.

The parasites encountered were crustacean copepods and isopods besides platyhelminthic monogenea. Four species of copepods, Ergasilus versicolor, E. Lizae, Caligus bonito and Caligus sp. were encountered in the gills of M. curema. The platyhelminthic monogenea was not taxonomically identified. Higher number of parasites was registered on M. curema, with an average of 32.3 during the rainy season, and with an average of 5.3 during the dry season. The gills and the skin of the hosts were the preferred areas of parasitic fixation. However, the isopods were found on the tongue in the buccal cavity besides the gill chambers of the hosts (Figure 1). This study registered for the first time the occurrence of parasites on marine fishes from the coastal waters of Rio Grande de Norte, Northeastern Brazil (Cavalcanti, 2002; Cavalcani et al., 2004).

Table 1. Number of fish species examined and number of fish parasitized from the coastal waters of RN, Northeatern Brazil, during a pioneering study conducted in 2001/2002

Family	Fish Species	Number of fish examined	Number of fish parasitized
Sciaenidae	Larimus breviceps Cuvier, 1830	6	0
	Menticirrhus americanus (Linnaeus, 1758)	75	3
	Cynoscion acoupa (Lacepede, 1801)	6	0
Mugilidae	Mugil curema Valencienners, 1836	31	19
Ephippidae	Chaetodipterus faber (Broussonet, 1782)	8	2
Haemulidae	Pomadasys corvinaedormis (Steindachner, 1868)	155	27
	Pomadasys crocro (Curier, 1830)	1	1
	Conodon nobilis (Linnaeus, 1758)	31	0
	Anisotremus virginicus (Linnaeus, 1758)	1	0
	Haemulon parra (Desmarest, 1823)	1	0
Ariidae	Bagre marinus (Mitchill, 1815)	2	1
Trichiuridae	Trichiurus lepturus Linnaeus, 1758	4	1

Contd...

Table 1 : Contd...

Clupeidae	*Pellona harroweri* (Fowler, 1917)	92	1
	Opisthonema oglinum (Lesueur, 1818)	100	1
Carangidae	*Caranx hippos* (Linnaeus, 1766)	2	1
	Selene setapinnis (Mitchill, 1815)	5	0
	Oligoplites saurus (Bloch & Schneider, 1801)	7	4
	Caranx latus Agassiz, 1831	1	0
	Trachinotus carolinus (Linnaeus, 1766)	6	1
	Chloroscombrus chrysurus (Linnaeus, 1766)	47	10
Polyemidae	*Polydactylus virginicus* (Linnaeus, 1758)	35	3
Gerreidae	*Diapterus rhombeus* (Cuvier, 1829)	18	1
Engraulidae	*Anchoviella lepidentostole* (Fowler, 1911)	1	0
	Anchovia clupeoides (Swainson, 1839)	28	2
Scombridae	*Scomberomorus brasiliensis* Collette, Russo & Zavala-Camin, 1978	10	8
Sphyraenidae	*Sphyraena barracuda* (Edwards, 1771)	3	1
Centropomidae	*Centropomus undecimalis* (Bloch, 1792)	1	1
Achiridae	*Achirus lineatus* (Linnaeus, 1758)	1	0
Lutjanidae	*Lutjanus synagris* (Linnaeus, 1758)	9	2
Total			
16 Families	**29 Fish Species**	**687**	**90**

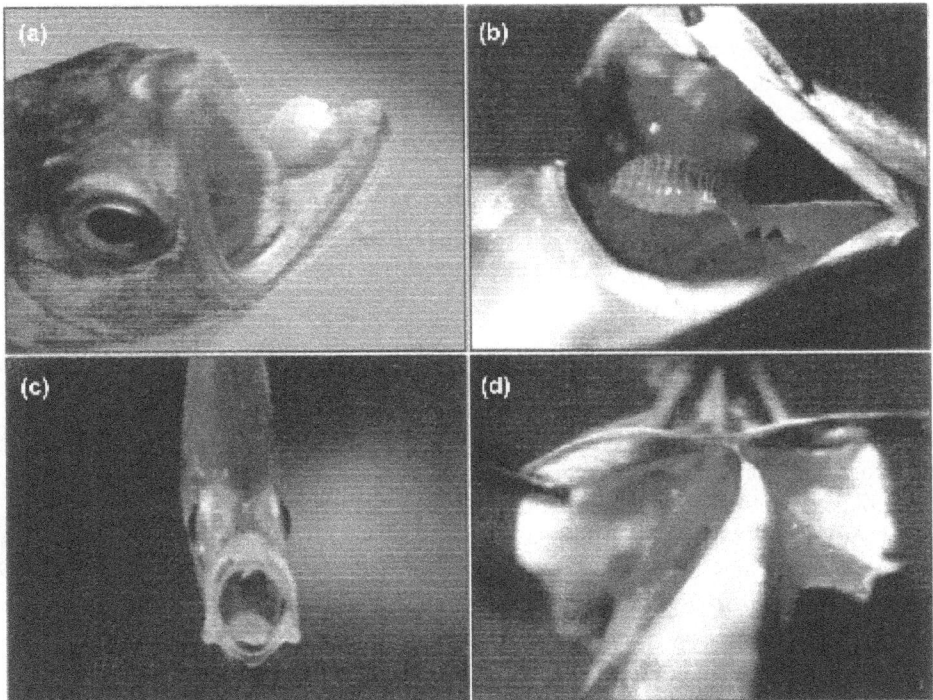

FIG. 1. Fish parasitized by isopods: (a) in the buccal cavity of *Oligoplites saurus;* (b) in the gill chamber of *Scomberomorus brasiliensis;* (c) and (d) in the buccal cavity and gill chambers of *Chloroscombrus chrysurus*. (Illustrations provided by Dr. E.T.S. Cavalcanti, Universidade Federal do Rio Grande do Norte, UFRN/RN, Brazil).

PARASITIC COPEPODS

The occurrence of two copepod parasites, *Ergasilus versicolor* and *E. lizae* (Copepoda: Ergasilidae) in the gill chambers of the white mullet, *Mugil curema* (Mugilidae) was reported (Cavalcanti et al., 2005). During the rainy season, 383 specimens of *E. versicolor* (Fig. 2A) were registered with a prevalence of 58.06%, and during the dry season 22 specimens of *E. lizae* were registered, with a prevalence of 3.23%. The parasites showed specificity in relation to male fish hosts, which was 66.67% during the rainy season and 100% during the dry season. This study was a first record of two parasitic copepod species of the family Ergasilidae on *M. curema* captured from the coastal waters of RN, Northeastern Brazil.

The presence of the copepod *Lernanthropus rathbuni* (Copepoda: Lernanthropidae) parasitizing Poma-dasys corvinaeformis (Osteicthyes, Haemulidae), in coastal waters of Rio Grande do Norte, Brazil was registered (Cavalcanti et al., 2006a). One hundred and fifty five specimens of this marine fish were collected from March, 2001 to June, 2002, out of which 107 fish samples were captured during the rainy season of the region (March-August) and 48 fish during the dry season (September-February). Under necropsy twenty-seven fish were found to be parasitized by *L. rathubni* (Fig. 2B), with a prevalence of 17.4% and mean infection intensity of 13.02. The preferred locations of fixation were the gills (97.56%) and to a lesser degree the tegument (2.44%) of the host fish. Maximum number of parasites occurred in female hosts during the rainy season.

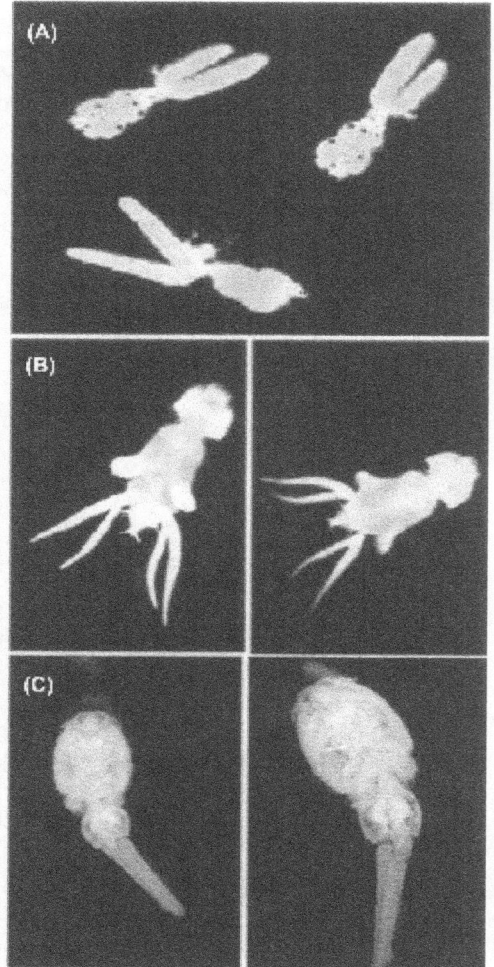

Fig. 2.   (A) Females of copepod parasite *Ergasilus versicolor* encountered in the gill chambers of *Mugil curema* (x100); (B) Ventral and dorsal views of male *Lernanthropus rathbuni* encountered in *Pomadasys corvinaeformis* (x100); (C) Ventral and dorsal views of female Caligus bonito with ovigerous sacs, encountered in Mugil curema (x200). (Illustrations 2(A), (B) and (C) were provided by Dr. E.T.S. Cavalcanti, Universidade Federal do Rio Grande do Norte, UFRN/RN, Brazil).

Two parasitic copepod species, *Caligus bonito* and *Caligus* sp. (Copepoda: Caligidae) (Fig. 2C) were reported parasitizing on the white mullet, *Mugil curema*. From the 31 fish samples captured form the coastal waters of Rio Grande do Norte, during 2001/2002, parasites occurred only on two fish samples. Four male specimens of C. bonito (66.7%), and two female specimens of *Caligus* sp. were registered. Prevalence values of 12.9% and 6.45% were recorded for *C. bonito* and *Caligus* sp., repectively. The preferred area of fixation by the parasites was the gill chambers of the hosts (Cavalcanti et al., 2006b).

PARASITIC ISOPODS

Two isopod parasites, *Livoneca redmanni* Leach (Isopoda: Cymothoidae) and Rocinela signata Schioedte & Meinert (Isopoda: Aegidae) were registered for the first time on Serra Spanish mackerel, *Scomberomorus brasiliensis* (Osteichthyes: Scombridae) captured from the coastal waters of Northeastern Brazil. *L. redmanni* was encountered both in the buccal cavity and the branchial chambers of the host fish, whereas *R. signata* was found only in the gill chambers (Fig. 3). These isopods

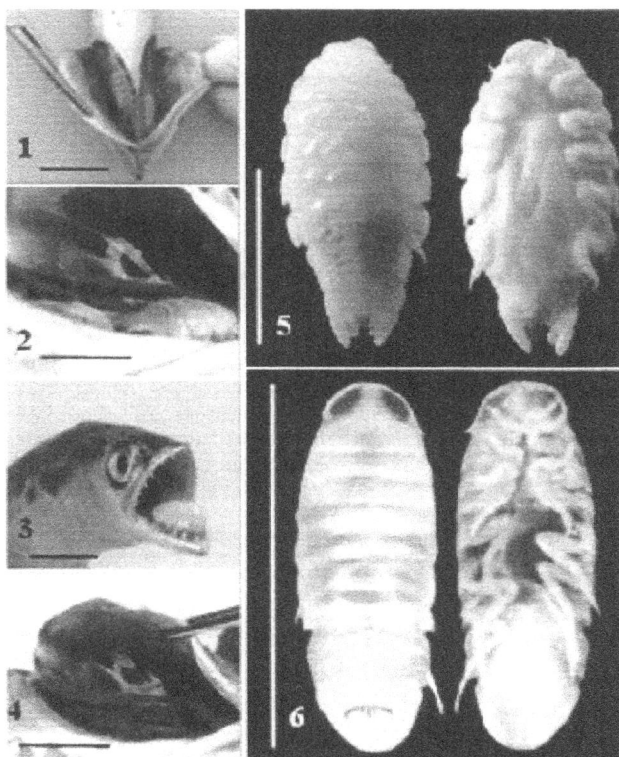

FIG. 3.   Isopod parasites *Livoneca redmanni* and *Rocinela signata* in *Scomberomorus brasiliensis.* (1-2) *L. redmanni* in the branchial chamber (scale = 20 mm); (3) *L. redmanni* in the buccal cavity (scale = 2 mm); (4) Damaged branchial arches of he host (scale = 10 mm); (6) Dorsal and ventral views of *Rocinela signata* (scale = 10 mm). (Illustrations provided by Dr. J. T. A. Ximenes-Lima, Universidade Federal Rural do Semi-Arido, UFERSA/RN, Brazil).

showed a preference for immature and maturing stages of the host fish (Lima, 2004; Lima et al., 2005). Reproductive stage of the female isopod *L. redmanni* parasitizing *S. brasiliensis* showed the presence of eggs and larve in the marsupial pouch (Fig. 4) (Lima, 2008).

R. signata has been reported as a fish parasite which is responsible for reducing their growth rates. This parasite was given the common name of "Monogram Isopod" because of the inverted W-shape mark on its pletelson, as shown in Fig. 3(6).

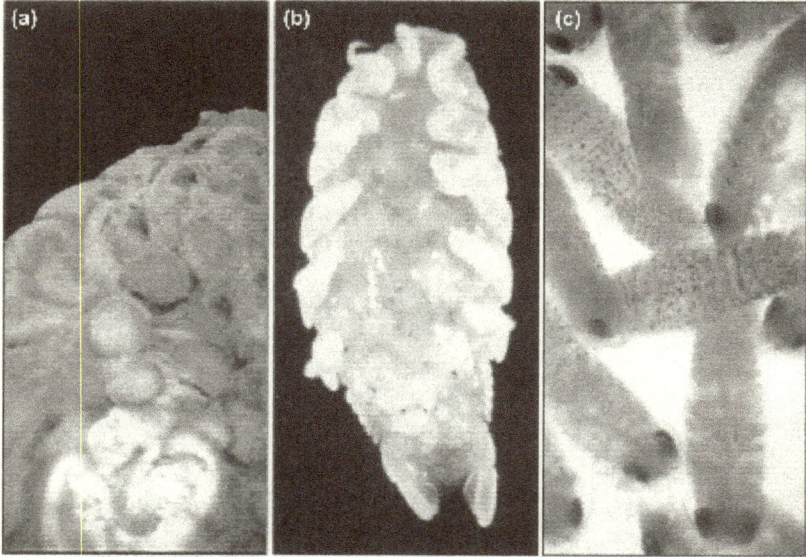

FIG. 4.   Reproductive stages of the female isopod parasite *Livonceca redmanni* in the host fish *Scomberomorus brasiliensis*: (a) Eggs in the marsupial pouch of the female (scale: 5 mm); (b) Black points seen on the ventral side are the eyes of larvae inside the marsupium of the female (scale: 5 mm); (c) larvae ready to swim (scale: 2 mm). Illustration provided by Dr. J. T. A. Ximenes-Lima, Universidade Federal Rural do Semi-Arido, UFERSA/RN, Brazil).

A new species *Cymothoa spinipalpa* sp. nov. (Isopoda: Cymothoidae)a buccal cavity parasite (Fig. 5) of the marine fish, *Oligoplites saurus* (Osteichthyes: Carangidae) from Rio Grande do Norte State, Brazil, was described and registered (Thatcher et al., 2007). In this new species, the anterior margin of the cephalon is doubled ventrally over the base of the antennae. In this respect, though it resembles *C. recifea* Thatcher & Fonseka, 2005 of Pernambuco State, Brazil, it differs in being much smaller with basal carinae on the pereopods 4 to 7. This new species was distinguished from all known *Cymothoa* spp. by the mandibular palps which are entirely covered with small spines in adult males. This paper described the ninth Brazilian species of *Cymothoa*.

The crustacean isopod parasite, *C. spinipalpa* sp. nov. was also encountered on the tongue in the buccal cavity of the marine fish, *Oligoplites palometa*. The parasitic indices recorded were high, 641% of prevalence, mean intensity of 2.02 parasites per host and abundance of 1.29 parasites per fish sampled. The heighted rates of parasite prevalence and abundance occurred in the dry season (Araujo, 2005; 2008).

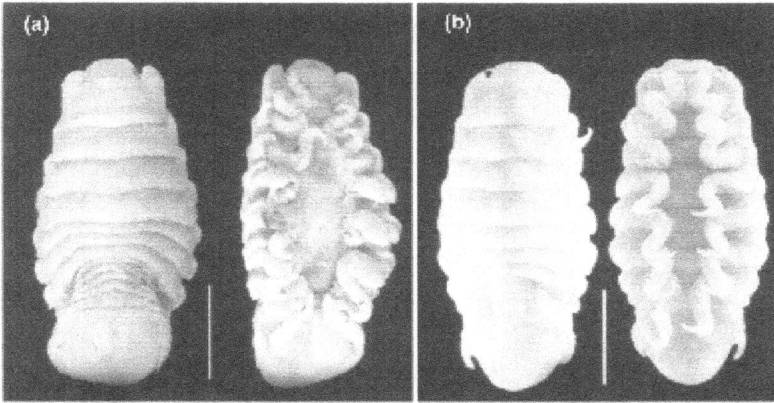

FIG. 5. Isopod *Cymothoa spinipalpa* found in the buccal cavity of *Oligoplites saurus* (a) Dorsal and ventral view of female; (b) Dorsal and ventral view of male (scale = 5 mm). (Illustration provided by Mr. G.S. Araujo, Universidade Federal do Rio Grande do Norte, UFRN/RN, Brazil).

The isopod parasites, *Livoneca redmanni* and *Cymothoa spinipalpa* (Isopoda: Cymothoidae) were registered on the host *Chloroscombrus chrysurus* (Osteichthyes, Carangidae) captured in the Northeastern coastal waters of Brazil (Fig. 6). During this study, the fish samples were netted on a monthly basis from January to December, 2006. The body surface, buccal cavity and branchial chambers of 221 fish samples were examined and 40 speciemens were found parasitized. On the hosts a total of 44 isopod parasites were collected, which included 14 specimens of *L. redmanni* and 30 of *C. spinipalpa*. All specimens of *L. redmanni* were found fixed in the branchial chambers. The males of *C. spinipalpa* did not show any preference for fixation site, but the females showed preference to the buccal cavity of the hosts. Isopod parasites did not cause any harmful effects on the body condition of the parasitized fishes. The relationship between the total length of sesx grouped *L. redmanni* and the total length of the females of *C. spinipalpa*, showed a positive correlation with the total length of the hosts. However, the males of *C. spinpalpa* showed a negative correlation. Infections by both *L. redmanni* and the femals of *C. spinpalpa* occur early in the life history of the host, consequently the parasite grows simultaneously with the host and enters the reproductive phase (Costa *et al.*, 2009).

Parasitism in the marine fish Atlantic bumper, *Chloroscombrus chrysurus* (Osteichthyes: Carangidae) by the cymothoid isopod *Cymothoa spinipalpa* (Fig. 6) and information pertaining to biological aspects were investigated (Costa, 2007; Costa et al., 2010a). A total of 30 adults of *C. spinipalpa* were found in 26 of the 204 fish sampled. A prevalence of 12.8% and mean intensity of 1.15 parasites per fish were registered. The isopods were found in the branchial chamber and buccal cavity of the host. However, the parasitized and non-parasitized fish did not show any significant differences in their mean body mass, body length and condition factor. A positive correlation was found between the total length of female isopods and total body length of the hosts. The parasitic isopod females were larger than the males and showed a preference for the buccal cavity. *C. chrysurus* was reported as a new host to the isopod parasite, *Livoneca redmanni* captured in the coastal waters of Rio Grande do Norte, Brazil (Costa et al., 2010bb).

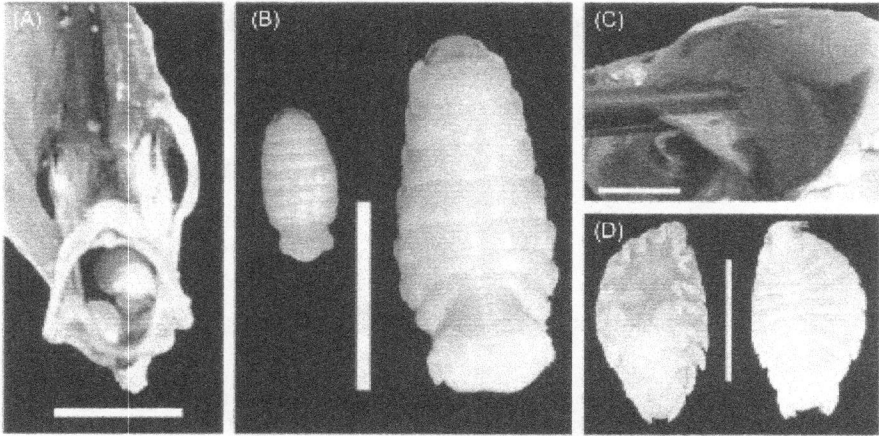

FIG. 6.   Isopod parasites *Livoneca redmanni* and *Cymothoa spinipalpa* found parasitizing the host *Chloroscombrus chrysurus*. (A) A pair of *C. spinipalpa* in the buccal c avity; (B) A dorsal view of a male and female of *C. spinipalpa*; (C) *L. redmanni* fixed on the branchial chamber; (D) Ventral and dorsal view of a female of *L. redmanni*. (scale bar = 10 mm). (Illustrations provided by Mr. E. F. S. Costa, Universitdade de Sao Paulo, USP/SP, Brazil).

The isopod *C. spinipalpa* does not exhibit parasitic specificity in relation to their hosts, since it parasitizes three different fish species, *C. chrysurus*, *Oligoplites saurus* and *O. palometa* of the family Carangidae. However, the higher parasitic indices were registered in *O. saurus* and *O. palometa*, with preference to the tongue of the host as the site of fixation. The parasitic indices were low for *C. chysurus*, and the parasites were found both in the buccal cavity and in the gill chambers of the hosts (Costa *et al.*, 2010c).

CO-EXISTENCE OF COPEPOD AND ISOPOD PARASITES

Mugil curema is a commercial fish species abundant in the Northeastern coastal waters of Brazil and the occurrence of crustacean ectoparasites on this fish was investigated during 2006 to 2007. It was found that *M. curema* was parasitized by ectoparasitic caligid copepod crustaceans, *Caligus bonito* and *Caligus* sp., ergasilid copepod crustaceans, *Ergasilus versicolor* and *E. lizae* besides isopod *Cymothoa spinipalpa*. Of the caligids detected, 66.66% was *C. bonito* an 33.33% was *Caligus* sp. *C. bonito* occurred on males of *M. curema* during the dry season and *Caligus* sp. occurred on female hosts during the rainy season. The prevalence of both caligid species was 3.23%. Of the ergasilids detected, *E. versicolor* (91.67%) occurred during the dry and rainy season, whereas *E. lizae* (8.33%) occurred only during the rainy season. Prevalence of *E. versicolor* was 35.48% and *E. lizae* was 3.23% . *C. spinipalpa* was detected during both seasons with a prevalence of 16.13%. All parasites preferred the branchial chambers as site of fixation (Cavalcanti, 2010; Cavalcanti et al., 2011).

The effects of parasites vary according to status and balance on the host-parasite relationships, where they can cause lesions ranging from low impact up to irreversible difficult situations. The presence of ectoparasites may effect the body parts of the host fish, causing gill filament atrophy, removal of brachial arcs, and obstruction of the mouth cavity leading to the destruction of the tongue, and sometimes leading to the death of the host (Rhode, 2005).

PARASITIC NEMATODS

First record of endoparasitism by the nematode, *Philometra* sp. (Nematoda: Philometridae) (Fig. 8) in lane snapper, *Lutjanus sysnagris* (Osteichthyes: Perciformes: Lutjanidae) was created fo the coast of Rio Grande do Norte, Brazil (Cavalcanti *et al.*, 2010). The parasites were found attached on the surface and in the ovarian tissue of mature female fish. The parasitic prevalence was 22.22%, mean intensity was 13 parasites per infected fish and mean density was 2.89 parasites per host sampled. the preferred location was the ovary, thus the parasite showed sex preference of host.

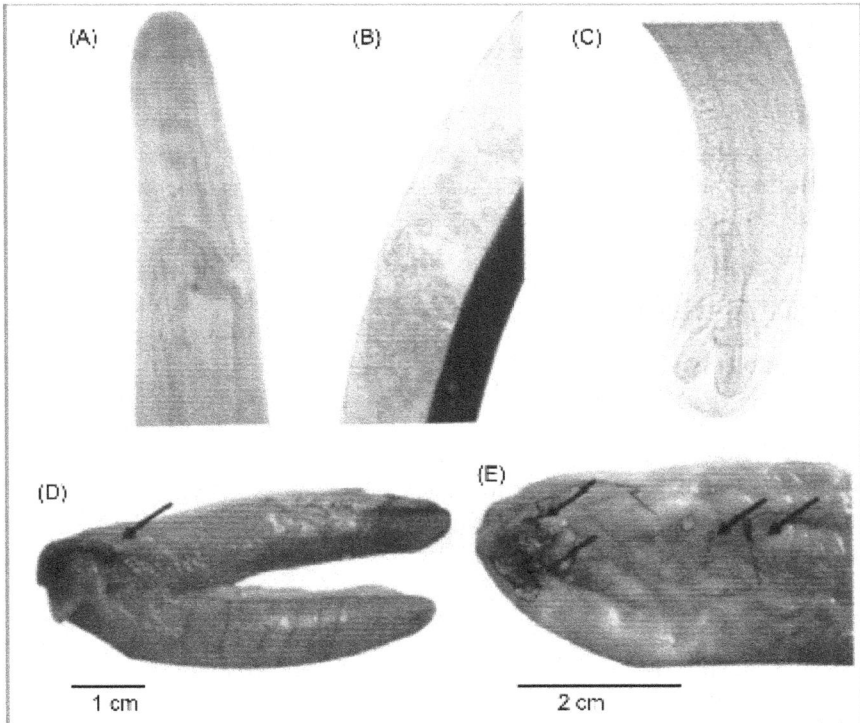

FIG. 7. Gravid female of *Philometra* sp: (A) anterior region; (B) middle region with eggs; (C) posterior region; (D) parasitized fish ovary; (E) dissected ovary of the fish *Lutjanus synagris* showing the endoparasite *Philometra* sp. (indicated by arrows). (Illustrations provided by Dr. E. T. S. Cavalcanti, Universidade Federal do Rio Grande do Norte, UFRN/RN, Brazil).

Gonad infection can lead to parasitic castration, depending on the exact location and intensity of infection. Parasites located in the ovarian or testicular stroma are less pathogenic than those located within oocytes or seminiferous tubules. In both sexes of the host, heavy infections may lead to parasitic castration. Infected seminiferous tubules contained few or no spermatids or spermatozoa and the parasite almost entirely replaced the contents of the ovaries (Sitja–Bobadilla, 2009).

FUTURE PERSPECTIVES

Marine fish parasites cause commercial losses in both the mariculture and fisheries industries and may involve human health and socio-economic implications. A full understanding of the diverse effects of fish parasites on their hosts is therefore important to the development and maintenance of fisheries.

Fish serve as hosts to a range of parasites which are taxonomically diverse and exhibit a wide variety of life cycle strategies. However, studies on marine fish parasites could still be considered scarce in comparison to the vast fish diversity in the coastal waters of Northeastern Brazil. Furthermore, the correct parasite identification is of very important, yet for all, the identification of many of fish parasites is still not clear.

Many of the consequences suffered by infected fish are associated, in one way or another, with altered behaviour. Studies on effects of parasites on fish behaviour, or more specifically, infection-associated changes in the behaviour of marine fishes could be carried out.

Future studies will determine the actual contribution of the fish parasites to marine biodiversity. Larger samples and long term studies are needed to describe and analyze the diversity of marine fish parasites and to verify their potential threats to the coastal fisheries.

ACKNOWLEDGEMENTS

The authors wish to thank the National Council for Scientific and Technological Development of Brazil (CNPq) for the financial support awarded during the study period (S. Chellappa, J. T. A. X. Lima, E.F.S. Costa, N. T. Chellappa) and the Federal Post-Graduate Agency of Brazil (CAPES/MEC) (Scholarships granted to E.T.S. Cavalcanti and G. S. Araujo). The authors wish to thank collegues, especially Mr. Wallace Silva do Nascimento, of the Ichthyology laboratory of UFRN, Brazil, for organizing the figures and many insightful discussions.

REFERENCES

Adlerstein, S. A. and Dorn, M. W. 1998. The effect of Kudoa paniformis infection on the reproductive effort of female pacific hake. Canadian Journal of Zoology, 76, 12, 2285 2289.

Araujo, G. S. (2008). Ecologia parasitaria de isopodos e biologia reprocutiva em tibiro, Oligoplites spp. (Osteichthyes: Carangidae) das aguas costeiras de Natal, Rio Grande do Norte. Masters Dissertation, Universidade Federal do Rio Grande do Norte, UFRN/RN, Brazil, p. 97.

Boxshall, G. A.; Montu, M. A. (1997). Copepoda parasitic on Brazilian coastal fishes: a handbook. Nauplius, v. 5, n.1, pp. 1-225.

Brusca, R. C. (1981). A monograph on the Isopoda: Cymothoidae (Crustacea) of the eastern Pacific. Zoological Journal of the Linnean Society, 73, 117-199.

Bunkley-Williams, L.; Williams Jr., E. H. (1998). Isopods associated with fishes: a synopsis and corrections. Journal of Parasitology, 84, 893-896.

Bunkley-Williams, L.; Williams Jr., E. H. and Bashirullah, A. K. M. (2006). Isopods (Isopoda: Aegidae, Cymothoidae, Gnathiidae) associated with Venezuelan marines fishes (Elasmobranchii, Actinopterygii). Revista Biologia Tropical (International Journal of Tropical Biology), 54 (Suppl. 3), 175-188.

Cavalcanti, E. T. S. (2002). Ectoparasitas de peixes marinhos de valor commercial: tainha Mugil curema e uro, Pomadasys corvinaeformis de Ponta Negra, Rio Grando do Norte. Masters Dissertation, Universidade Federal do Rio Grande do Norte, UFRN/ RN, Brazil, p. 85.

Cavalcanti, E. T. S. (2010). Parasitos dos peixes marinhos de valor commercial no litoral do Rio Grande do Norte. Doctoral Thesis, Universidade Federal Rural de Pernambuco, UFRPE/PE, Brazil, p. 130.

Cavalcanti, E. T. S.; Pavanelli, G. C.; Chellappa, S. and Takemoto, R. M. (2004). Comunidade de metazoarios ectoparasitas de peixes de aguas costeiras de Ponta Negra, Natal, RN, Brasil. Ecologia Aquatica Tropical. Editores: N. T. Chellappa, S. Chellappa, J. Z. O. Passavante. Editora. ServGraf, Natal, RN. 157-165.

Cavalcanti, E. T. S.; Pavanelli, G. C.; Chellappa, S. and Takemoto, R. M. (2005). Ocorrencia de Ergasilus versicolor e E. lizae (Copepoda: Ergasilidae) na tainha, Mugil curema (Osteichthyes: Mugilidae) em Ponta Negra, Natal, Rio Grande do Norte. Arquivos de Ciencias do Mar. 38, 131-134.

Cavalcanti, E. T. S.; Chellappa, S.; Pavanelli, G. C. and Takemoto, R. M. (2006a). Presenca de Lernanthropus rathuni (Copepoda: Lernanthropidae) no coro, Pomadasys corvinaeformis (Osteichthyes, Haemulidae) em aguas costeiras do Rio Grande do Norte. Arquivos de Ciencias do Mar. 39, 134-137.

Cavalcanti, E. T. S.; Chellappa, S.; Pavanelli, G. C. and Takemoto, R. M. (2006b). Registro de ocorrencia de Caligus bonito e Caligus sp. (Copepoda: Caligidae) na tainha, Mugil curema (Osteichthyes, Mugilidae), no litoral de Natal, Rio Grande do Norte. Arquivos de Ciencias do Mar. 39, 131-133.

Cavalcanti, E. T. S.; Takemoto, R. M.; Alves, L. C. and Chellappa, S. (2010). First record of endoparasite Philometra sp. (Nematoda: Philometridae) in lane snapper, Lutjanus synagris from the coast of Rio Grande do Norte, Brazil. Marine Biodiversity Records, Cambridge University Press, UK. 3(3), 1-4.

Cavalcanti, E. T. S.; Takemoto, R. M.; Alves, L. C.; Chellappa, S. and Pavanelli, G. C. (2011). Ectoparasitic crustaceans on mullet, Mugil curema (Osteichthyes: Mugilidate) in the coastal waters of Rio Grande do Norte, Brazil. Acta Scientiarum (Biological Sciences), 33, 3, 357-362.

Costa, E. F. S.; (2007). Parasitas isopodes encontrados nos peixes marinhos das aguas costeiras do RN. Graduate Monograph, Universidade Federal do Rio Grande do Norte, UFRN/RN, Brazil. 74p.

Costa, E. F. S.; Oliveira, M. R. and Chellappa, S. (2009). Isopod parasites of the fish Chloroscombrus chrysurus (Osteichthyes, Carangidae) in the Northeastern coastal waters of Brazil. Proceedings of XVI Congresso Brasileiro de Engenharia de Pesca. Natal/RN/Brasil. 907-913.

Costa, E. F. S., Oliveira, M. R. and Chellappa, S. (2010a). First record of Cymothoa spinipalpa (Isopoda: Cymothoidae) Parasitizing the marine fish Atlanic bumper, Chloroscombrus chrysurus (Osteichthyes: Carangidae) from Brazil. Marine Biodiversity Records, Cambridge) from Brazil. Marine Biodiversity Records, Cambridge University Press, UK. 3, 1, 1-6.

Costa, E. F. S. and Chellappa, S. (2010b). New host record for Livoneca redmanni (Leach, 1818) (Isopoda, Cymothoidae) in the Brazilian coastal waters with aspects of host-parasite interaction. Brazilian Journal of Oceaography, 58 (special issue IICBBM, 73-77.

Costa, E. F. S.; Oliveira, M. R.; Araujo, G. S. and Chellappa, S. (2010c). Parasito Cymothoa spinipalpa (Isopoda: Cymothoidae) de tres species de peixes (Osteichthyes: Carangidae) das aguas costeiras do Rio Grande do Norte, Brazil. Proceedings of SBPC 62. Natal, RN, Brazil. 1-2.

Klimpel, S., Seehagen, A., Palm, H. –W. and Rosenthal, H. (2001). Deep-water Metazoan Fish Parasites of the World. Logos Verlag, Berlin, p. 1316.

Lima, J. T. A. X. (2004). Biologia Reprodutivae e Parasitismo por Isopodes do Serra, Scomberomorus brasiliensis (Collette, Russo & Zavala-Camin, 1978) (Osteichthyes: Scombridae) no litoral do Rio Grande do Norte. Masters Dissertation, Universidade Federal do Rio Grande do Norte, UFRN/RN, Brazil, p. 153.

Lima, J. T. A. X. (2008). Estrategias reprodutivas e parasitarias de quarto especies de peixes das aguas costeiras do Sudoeste do Oceano Atlantico, Brasil. Doctoral Thesis, Universidade Federal do Rio Grande do Norte, UFRN/RN, Brazil, p. 157.

Lima, J. T. A. X.; Chellappa, S. and Thatcher, V. E. (2005). *Livoneca redmanni* Leach (Isopoda, Cymothoidae) e Rocinela signata Schioedte & Meinert (Isopoda, Aegidae), ectoparasitos de Scomberomorus brasiliensis Collette, Russo & Zavala-Camin (Osteichthyes, Scombridae) no Rio Grande do Norte, Brasil. Revista Brasileira de Zoologia. 22, 4, 1104-1108.

Palm, H. W. (2007). The concept of cumulative parasite evolution in marine fish parasites. Proceedings of the IV All-Russian Workshop on Theoretical and Marine Parsitology, Kaliningrad. 164-170.

Ravichandran, S.; Rameshkumar, G. and Balasubramanian, T. (2010). Infestation of isopod parasites in commercial marine fishes. Journal of Parasitic Diseases. 34, 2, 97-98.

Rohde, K. (Ed) (2005). Marine parasitology. University of New Englaand: CSIRO Publishing, p. 592.

Sitja- Bobadilla, A. (2009). Can Myxosporean parasites compromise fish and amphibian reproduction? Proceedings of the Royal Society, Biological Sciences, 276, 2861-2870.

Thatcher, V. E.; Araujo, G. S.; Lima, J. T. A. X. and Chellappa, S. (2007). *Cymothoa spinipalpa* sp. nov. (Isopoda, Cymothoidae) a buccal cavity parasite of the marine fish, Oligoplites saurus (Bloch & Schneider) (Osteichthyes: Carangidae) of Rio Grande do Nort Stat, Brazil. Revista Brasileira de Zoologia. 24, 1, 238-245.

Williams, E. H. Jr., Bunkley-Williams, L. (1996). Parasites off shore, big game sport fishes of Puerto Rico and the Western North Atlantic. Puerto Rico Department of Natural and Environmental Resources, San Juan PR, and the University of Puerto Rico, Mayaguez, p. 382.

□□□

12

ON GENUS *MYXIDIUM;* BUTSCHLI 1882
AND A NEW SPECIES *MYXIDIUM CHOLAI*
FROM *PUNTIUS CHOLA* (HAMILTON)

Kanwar Narain, M.K. Raina and P.L. Kaul

INTRODUCTION

The Genus *Myxidium* was established by Butschli, 1882 with *M. Leiberkuhni* Butschli, 1882 as its type species and the present work is a report of the genus form the gall bladder of the fish *Puntius Chola* (Hamilton) from Jammu.

Material and Methods

The fishes collected were pithed, gall bladder removed, cut open and kept in 0.65% saline. Very dilute solution of $CuSO_4$ (Jahn, Bovee and Jahn, 1979) was used to impede movements of protozoans. Vital and supravital dyes-toluidine blue and methyl blue were used for studying the protozoans.

Rapid fixation and staining techniques were used to show nuclei and other structures for protzoans in temporary preparation by Lugol's Iodine solution (C.F. Mackinnon and Hawes, 1961) and methyl green (Mackinnon and Hawes, 1961).

Myxosporidan spores were directly smeared on grease free slides and the smears fixed in Schaudinn's fixative for 15-30 minutes in Delafield's haematoxylin, differentiated in acid water; partially dehydrated in 70% ethanol, cleared in Xylene and mounted in Canada balsam.

Giemsas stain (BDH) was also used for staining MYXOPORIDIAN spores, but prior to staining the spores were treated with saturated aqueous solution of urea or 4% KOH solution for extrusion of polar filaments.

Stained specimens were studied under Meopta research microscope with maximum lens combination of 20x eye piece and oil immersion objective.

Living organisms were studied with the aid of Olympus phase contrast microscope.

Drawings were made to scale with the help of prism type camera lucida.

Genus Myxidium

The genus *Myxidium* was established by Butschli (1882) with *M. leiberkuhni* Butschli, 1882 as its type species. Lutz (1889 of. Kudo. 1966) created a new genus Cystodiscus for a myxosporidan which he recovered from gall bladder of Brazilian amphibian. However, this genus failed to gain acceptance (Kudo, 1996). Subsequently, numerous authors contributed towards different aspects of biology of the genus Myxidium such as Cohn (1896), Debaisieux (1918 and 1925), Cordero (1919), Kudo (1921 and 1943), Bremer (1922) and Naville (1930). Thelohan (1895), Auverbach (1910), Kudo (1919 and 1934), Shulman & Stein (1962) and Shulman (1966) published monographic accounts on myxosporidans including the genus Myxidium. Numerous species of genus Myxidium have been reported from all parts of the world in marine and freshwater fishes and rarely in amphibians and reptiles by Cepede (1906), Parisi (1912), Davis (1918 and 1947), Kudo (1920), Bond (1937 and 1938), Meglitsch (1937), Fantham, Porter and Richardson (1939 & 1940), Kudo and Sprague (1940), Noble (1943),Otto and Jahn (1943), Rice and Jahn (1943), Dogiel and Bogolepova (1975 and 1980), Dubina and Isakov (1976), Delvinquier (1987) and Fomena and Bouix (1987).

Slukhai (1975) worked on the systematic of some species of Myxidium of Cyprinidae. *Myxidium zealandicum* was established as a new species by Hine (1975). Later, he in the year 1979 studiedd its morphological variations and in 1980 he reviewed all the known species of Myxidium from eels. Ventura and Paperna (1985) studied the histopathological aspects of infection in European eel, Anguilla Anguilla in Portugal due to *Myxidium giardi* infection. Paperna, Hartley and Cross (1987) studied the ultrastructural aspects of trophozoites of M. *giardi* with special attention to their mode of attachment to the epithelium of the urinary bladder in Anguilla Anguilla.

In india, the genus Myxidium was reported for the first time by Bosanquet in 1910. He reported *Myxidium mackiei* from the kidneys of tortoise, *Trionyx gangeticus* Cuvier from Bombay. However, Ray (1933 a and b) was first to report the genus Myxidium from fish hosts, namely, *Clarias batrachus. Channa punctatus* and *Saccobranchus follilis*. Subsequently, Chakravarty (1939 and 1943), Lalitha Kumari (1969), Choudhury and Nandi (1973), Bajpai & Haldar (1982), Sarkar (1982, 1985a and 1986a), Sarkar, Mazumdar and Pramanik (1985) and Sarkar and Choudury (1986) described numerous new species of the genus Myxidium from Indian region.

Myxidium cholai n. sp. (Fig. 1a-f)

Type Host : *Puntius chola* (Himilton), *Barbus tetrarupagus*, Day from Ghoomanasa Nallah Jummu.

PREVALENCE

1 out of 15 fish examined was infected.

DESCRIPTION

Myxidium cholai n. sp. is coelozoic and the mature spores are found free floating in the gall bladder.

SPORES

The spores are broadly lenticular in front view (= valvular view) with of the sides markedly convex (Fig. 4ab). In side view (= sutural view) the spores are elongated, cylindrical and more or less sigmoid in shape with slightly pointed ends (Fig. 4c-e).

The spores measure 9 - 14.5 (11.37 ± 1.68) μm x 6 – 6.5 (6.3 ± 0.28) μm. The spore are 4.5 – 5.5 (4.75 ± 0.43) μm thick. The sutural line is faintly visible and somewhat curved. However, the sutural ridge is not discernible. No striations were seen on the spore valves. The polar capsules are two in number, equal in size and located at opposite poles and are broadly pysiform. The polar capsules measure 2-4 (2.39 ± 0.32) 4m × 2–3 (2.35 ± 0.11) μm. The coils of the polar filament are not visible in fresh spores. However, when extruded the polar filaments are clearly visible as hyaline thread-like structures and measure 10-18 (13.5 ± 2.12) μm. The sporoplasm is situated between the two polar capsules and is finely granular in consistency. Two sporoplasmic nuclei are visible in fresh as well as Giemsa stained preparations Fig. 1a-f

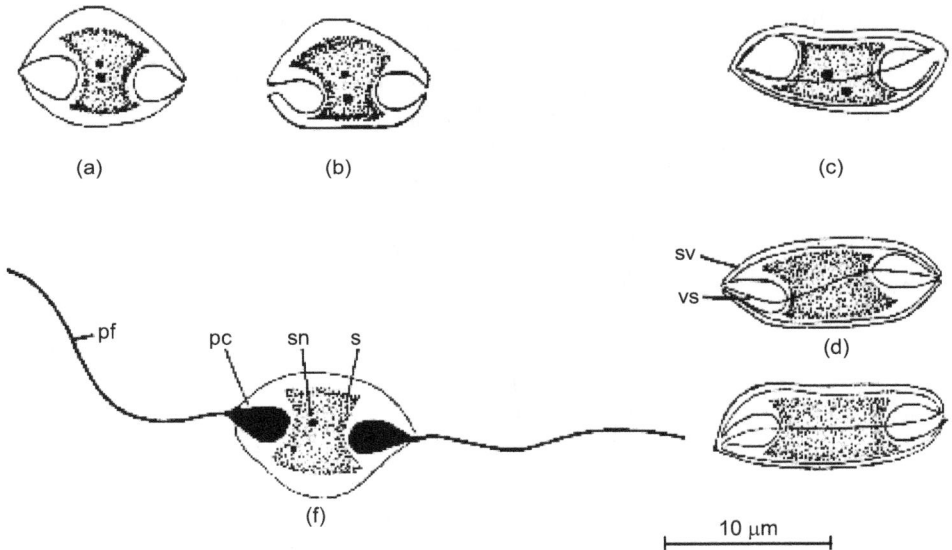

(a)   (b)   (c)

(d)

(f)   10 μm

FIG. 1(a-f):   Line drawings of the spores of *Myxidium cholai* n. sp.: (a&b) Fresh spores in front view, (c–e) Fresh spores in side (= sutural) view, (f) Giemsa stained spore with extruded Polar filament.

DISCUSSION

Myxosporidan taxonomy including Myxidium Butschli, 1882, is primarily based on the following features: (i) the spore morphology, particularly its overall shape and size, (ii) number, position, shape and size of polar capsules (= cnidocysts) and (iii) other spore characteristics such as presence or absence of valvular indentations and surface thickenings. These characteristics are fairly constant and consistent (Kudo, 1919 and 1934; Shulman and Stein, 1962; Shulman, 1966; Mitchell, 1967 and 1977 and Lom, 1969a). Later workers, like, Bajpai and Haldar (1982) have also supported this view and they state that, "the most important taxonomic character of myxosporidans is the infective stage i.e. the spore". Some authors also consider the number of coils and the angle of coiling of polar filament inside the polar capsule as taxonomically useful character at specific level (Lom, 1969). Although the length of the extruded polar filament is not considered to be a valuable character from taxonomical point of view by Mitchell (1977), according to whom the extruded polar filament is of unknown variability ans significance, yet many authors including Mitchell (1967) have given the length of polar filaments while describing their species. In the present investigation on a lot of variation was found in the length of polar filaments and therefore, the present work also confirms the validity of Mitchell's (1977) view. Bajpai and Haldar (1982) " have observed the simultaneous occurrence of two types of spores, one having a small pair and the other having large pair of polar filaments". But, they have not been able to give reasons for such variations. In the present study a clear cut distinction into such type of dimorphic spores was not observed but a significant variations in the length of the polar filaments was recorded within a particular range.

Besides spore morphology and morphometrics, other characteristics such as difference in the host species and site of infection have also been used for differentiating species (Hoffman *et al.*, 1965). Bond (1937) by using cross infection experiments showed that certain Myxoxporidea are host specific. Bond (1937), was also supported by other workers such as Mitchell (1967), according to whom some species of Myxosporidea, "are hardly distinguishable without knowledge of the hosts and / or tissues they infect. Tissue specificity is indicated for nearly all species." On the other hand the works of some authors, like, Mukherjee and Haldar (1981) and Bajpai and Haldar (1982) have shown that host and tissue specificity seem to be highly variable among species. Keeping all the above important features in view the relationship of present species with other known species is discussed.

On comparison with the known myxosporidans the present form was found to be congeneric with Myxidium Butschli, 1882 because the spores are fusiform and possess two polar capsules at the opposite ends.

A perusal of the literature reveals that 13 species of Myxidium have been reported from Indian fishes. These are:

1. *Myxidium* sp. Ray, 1933.
2. *M. glossogobii* Chakravarty, 1939.
3. *M. heteropneusti;* Chakravarty, 1943.
4. *M. lieberkuhni* Butschli, 1882 (first reported in India by Chakravarty, 1943).
5. *M. procerum* var. *calceriferi* Chakravarty, 1943.
6. *M. aori* Lalitha Kumari, 1969.
7. *M. boddaerti* Chouhury & Nandi, 1973.
8. *M. apocryptae* Bajpai & Haldar, 1982.
9. *M. striatusi* Sarkar, 1982.
10. *M. fasciatum* Sarkar, 1985
11. *M. islampurium* Sarkar, Mazumdar and Pramanik, 1985.
12. *M. mystusium* Sarkar, 1986.
13. *M. sciaenae* Sarkar, 1986.

Therefore, in view of the above differences the present form is given a separate specific status. This is the first report of the genus from Jammu region.

Table 1. Showing the comparative measurements (in microns), sites of infection and hosts 14 Species of *Myxidium* Butschli, 1882 reported so far from the freshwater fish of India.

Structure	M. sp. Mean	Range	M. glossogobii Mean	Range	M. heteropneusti Mean	Range
Length of the spore	-	-	(12-15)		14.42	-
Breadth of the spore	-	-	(8.5-10)		6.18	-
Length of the polar capsule	-	-	(3.1-4.1)		-	(4.1-6.1)
Breadth of the polar capsule	-	-			4.1	-
Infection locus	Gall bladder		Gall bladder		Gall bladder	
Host/s	Clarias batrachus, Heteropneustes fossilis (= Saccobranchus fossilis) and Channa punctatus (= Ophiocephalus punctatus)		(Hamilton) (= Gobius giuris)	Glossogobius giuris	Heteropneustes fossilis (= Saccobranchus fossilis)	
Refernce	Ray, 1933		Chakravarty, 1939		Chakravarty, 1943	

Structure	M. mikundae (=M. lieberkuhni) Mean	Range	M. procerum var. calceriferi Mean	Range	M. aori Mean	Range
Length of the spore	14.1	(12.4-15.0)	-	(23-27)	-	(11.4-13.6)
Breadth of the spore	3.6	(4.1-5.0)	6.18	-	-	(5.4-7.1)
Length of the polar capsule	4.1	-	8.2	-	-	(3.6-4.3)
Breadth of the polar capsule	2.0	-	4.1	-	-	(2.1-2.9)
Infection locus	Gall bladder		Gall bladder		Gall bladder	
Host/s	Anabas testudineus		Lates calcerifer		Macrones dor (= Mystus aor)	
Reference	Chakravarty, 1943		Chakravarty, 1943		Lalitha Kumari, 1969	

Contd...

Table 1 : Contd...

Structure	M. boddaerti Mean	Range	M. apocryptae Mean	Range	M. striatusi Mean	Range
Length of the spore	15.48	-	34.5	(30.7-38.2)	14.53	(11.1-18.7)
Breadth of the spore	7.7	-	5.1	(4.2-6.6)	5.61	(4.7-7.0)
Length of the polar capsule	-	-	11	(9.1-12.5)	4.47	(3.7-5.6)
Breadth of the polar capsule	-	-	2.5	-	3.04	(2.8-3.7)
Infection locus	Gut contents		Gall bladder		Gall bladder	
Host/s	Boleopthalmus boddarti (= B. boddarti)		Apocryptes bato		Channa striatus (= Ophicephalus striatus)	
Reference	Choudhury & Nandi, 1973		Bajpai & Haldar, 1982		Sarkar, 1982	

Structure	M. fasciatum Mean	Range	M. islampurium Mean	Range	M. mystusium Mean	Range
Length of the spore	16.0	(14.4-17.6)	9.93	(8.5-12.0)	13.08	(11.0-16.65)
Breadth of the spore	5.9	(5.6-6.4)	3.92	(3.0-6.0)	5.89	(5.0-7.49)
Length of the polar capsule	4.5	(4.0-4.8)	3.59	(3.0-4.5)	4.83	-
Breadth of the polar capsule	4.0	(3.2-4.8)	2.89	(2.0-3.0)	4.5	-
Infection locus	Gall bladder		Gall bladder		Gall bladder	
Host/s	Colisa fasciatus (= Trichogaster fasciatus)		Channa marulius		Mystus vittatus	
Reference	Sarkar, 1985		Sarkar, Mazumdar & Pramanik, 1985		Sarkar, 1986	

Structure	M. cholai n. sp. Mean	Range
Length of the spore	11.37	(9-14.5)
Breadth of the spore	6.3	(6.0-6.5)
Length of the polar capsule	3.39	(3.0-4.0)
Breadth of the polar capsule	2.35	(2.0-3.0)
Infection locus	Gall bladder	
Host/s	Puntius chola (= Barbus tetarupagus)	
Reference	Present work	

REFERENCES

AKHMEROV, A. K., (1960). Myxosporidia of fishes from the Amur river basin, Rybm. *Khoz. Nutr. Vod. LatSSR*, 5: 240-307.

ANTYCHOWICZ, J and ROGULSKA, A. (1985), Investigations on the control of *Ichthyophthirius multifiliis* in the crap. *Med. Water*, 41 (5): 269-272.

AYERBACH, M. 1910. *Die Cnidosporidien*. Leipzig. p. 261.

BAJPAI, RRN and HALDAR, D. P. (1982). Observations on tw. Myxosporidan parasites (MYXOZOA : MYXOSPOREA) from the gall bladder of fresh water fishes. Arch. Protistenk. 125 : 129-136.

BAUER, O.N. MUSSELIUS, U. A. and STRELKOV, Y. A. (1973) *Diseases of Pond Fishes* (In Russian. English translation by A. Mercado and edited by O. Theodor) Israel Program for scientific translations Ltd. Jerusalem, p. 220.

BOND, F. F., (1937). Host specificity of the Myxosporidia of *Fundulus heteroclitus* (Linn.). *J. Parasit.* 23 : 540-542.

BOSANQUET, W. C., (1910). Brief notes on two myxosporidian Organisms (*Pleistophora hippoglossoides* n. sp and *Myxidium mackiei* n. sp.) *Zool. Anz.*, 35 ;434-438.

BREMER, H. (1992). Studien Uber Kernbau and Kernteilung. Von *MYXIDIUM lieberkuhni* Butschli *Arch Protist.*, 45 : 273

BUTSCHLI, O. (1882), Myxosporidia. Bronn's klass.. *Ordn. Protozoa* 1 , pp. 590-603.

CANEL'LA, M.F. (1972), Contributions a la connaissance des Cilies. VII. Ce qu'on ne connait pas sur un ho

CEPEDE, M. C., (1906). Myxidium giardi cepede, et la pretendue munite des Anguilles a l'egard des infections myxosporidienne. *Seances Soc. Biol. Paris.* 60 : 170-173.

CHAKRAVARTY, M. M, (1939). Studies on Myxosporidia from the fishes of Bengal with a note on myxosporidian infection in aquaria fishes. *Arch. Protistenik*, 92 : 169 – 178.

CHAKRAVARTY, M. M., (1943). Studies on Myxosporidia from the Common food fishes of Bengal *Proc. Indian Acad. Sci.* (B), 18 : 21- 35.

CHAKRAVARTY, M. and BASU, S. P. (1948). Observations on some Myxosporidia in fishes with an account of nuclear Cycles in one of them. *Proc. Zool. Soc. Bengal*, 1 : 23-33

CHEN, C. L. (1955). The Protozoan parasites from four species of Chinese pond fishes: *Ctenopharyngodon idella, Mylopharyngodon piceus Aristichthys nobilis* and *Hypophthalmichthys molitrix. Acta Hydrobiol Sinica*, 1 : 123-164.

CHOUDHURY, A and NANDI, N. C. (1973). Studies on Myxosporidian parasites (Protozoa) from an esturine gobiid fish of west Bengal. *Proc. Zool. Soc.* Calcutta, 26 : 45-55.

COHN, L. (1896). Uber die Myxosporidien Von *Esox Lucius* and *Perca fluviatilis. Zool. Jber. Abt. Morph.* IX PP. 227-272

CORDERO, E. H. (1919). *Cystodiscus Immersus* Lutz : Mixoxporidio de los Datracios del Uruguay. Physis, 4 : 403.

DAVIS, H. S. (1918). The Myxosporidia of the Beaufort Region – A Systematic and biologic study. *Bull. Bur. Fish.*, Wash., 35 : 199-244.

1947. care and diseass of trout. *U. S. F. W. S. Res.* Rep. No. 12, p. 98.

1947a, Studies on the Protozoan Parasite of fresh water fishes. *U. S. Fish* Wild. *Serv. Fish. Bull.*, 51 : 1-29.

DEBAISIEUX, P. (1918). Notes sur le *Myxidium lieberkiihni. La* Cellule, 30 : 281.

1925. Etudes sur les Myxosporidies. III. *Arch. Zool. Exper. Gen.*, 64 : 353.

DELVINQUIER, B. L. J. (1986) (recd. 1987). *Myxidium immersum* (Protozoa : Myxosporea) of the Cane toad. *Bufo marines*. In Australian Anura, with a synopsis of the genus in amphibians *Aust. J. Zool.*, 34 (6) : 843-854.

DOGIEL, V. A. & BOGOLEPOVA, I. I. (1957). Parasites of the Baikal fishes. *Trud. Baik. Limn. Sta.*, 15 : 427-464.

DUBINA, V. R. and ISA kov, L. S. (1976) New species of Myxoxporidians from the gall bladder of bathial fishes. *Parazitologiya*, u (6) : 556-560.

FANTHAM, H. B., PORTER, A. and RICHARDSON, L. R. (1939) some Myxosporidia ound in Certain freshwater fishes in Quebec province Canada. *Parasitology*, 31 : 1-77.

1940. Some more Myxosporidia observed in Canadian fishes. *Parasitology*, 32 : 333-353.

FOMENA, A and BOUIX, G. (1986) (Recd. 1987) Contribution to the study of Myxosporidia of fresh water fishes in Cameroom : 1. New species of *Myxidium* Butschli, 1882. *Acta Trop.*, 43 (4) : 319-334.

GOPALAKRISHAN, V. (1964). Recent development in the prevention and control of parasites of fishes Cultured in Indian waters. *Proc. Zool. Soc.*, Calcutta, 17 : 95-100.

HINE, P. M. (1975). Three new speices of Myxidium (Protozoa : Myxosporidia) parasitic in *Anguilla australis* Richardson, 1848 and *A. dieffenbachia* GRAY, 1882 *Journal R. Soc. N. W.*, 5 (2) : 153-161.

HINE, P. M. (1979). Factors affecting the size of spores of *Myxidium zealandicum* Hine, 1995 (Protozoa : Myxosporida). *Newzealand J. mar. fresh W. Res.* 13 (2) : 215-223.

HINE, P. M. (1980). A review of some species of *Myxidium* Butschli, 1882 (Myxozosporea) from eels (*Anguilla* spp.). *Journal Protozool.*, 27 (3) : 260-267.

HOFFMAN, G. L. (1967) *Parasites of North American Freshwater Fishes.* Univ. of California Press, Berkeley.

HOFFMAN, G. L. (1978). Ciliates of fresh water fishes. In : *Parasitic Protozoa Vol. II* (J. P. Kreier, ed.) Academic Press, London. pp. 583-632.

HYMAN, L. H. (1940). *The Invertebrates : Protozoa through Ctenophora* Vol I Mc Graw- Hill, New York p. 726.

KUDO, R. R. (1919). Studies on Myxosporidia. A synopsis of genera and species. III. *Biol. Monogr.*, 5 : 241-503.

KUDO, R. R. (1920). Cnidosporidia in the vicinity of Urbanans. *III Acad. Sci.*; 18 : 298-303.

KUDO, R. R. (1921). On some Protozoa parasitic in fresh water fishes of New York *J. Parasit.*, 7 : 166-174.

KUDO, R. R.(1934). Studies on some Protozoan prarsites of fishes of Illinois. *III Biol. Monogr.*, 13 : 7-44.

KUDO, R. R. (1943). Further Observations on the Protozoan Myxidium *serotinum*, inhabiting the gall bladder of North American Salientia. *J. Morphol.*, 72 : 263-277.

KUDO, R.R. (1966). *Protozoology.* 5th ed. Charles C Thonias spring field , Illinois, pp. 1174.

KUDO, R. R. and SPRAGUE, V. (1940). On *Myxidium immersum* (Lutz) and *M. serotinum n.* sp., two myxosporidian parasites of Salientia of South and North America. *Revista Med. Trop. Parasit. Bacteriol. Clin Lab.*, 6 : 65-73.

LALITHA KUMARI, P.S. (1969). Studies on parasitic Protozoa (Myxosporidia) of fresh water fishes of Andhra Pradesh, India. *Riv. Parasit.*, 30 : 153-226.

LOM, J. (1969). On a new taxonomic Character in Myxosporidia as demonstrated in despcription of two new species of *Myxobolus. Folia Parasit.*, 16 : 97-103.

LUTZ, A. (1889) Ueber eine Myxosporidium aus der Gallenlilase Brazilianischer Batrachier (*Cystodiscus immerses*). *Centra Ibl. Bakt. Parasit.*, 5 : 84-88.

MACKINNON, D. L and HAWES, R. S. J. (1961). Methods in Protology In : *An Introduction to the study of Protozoa* OXFORD UNIVERSITY PRESS. 384-446.

MAJEED. S. K., GOPINATH, C. and JOLLY, D. W. (1984). An Outbreak of white spot disease (*Ichthyophthirius multifillis*) in young fingerling rainbow trout (*Salmo gairdneri*).

J. Small. Anim. Pract., 25 (8) : 517-524.

MEGLITSCH, P. A. (1937). On some new and known Myxosporidia of the fishes of Illinois. u. *Parasit.*, 23 : 467-477.

MEGLITSCH, P. A. (1960). Some Coelozoic Myxosporidia from New Zealand fishes. 1. General and family Ceratomysidae. *Trans. Roy. Soc. N. Z.*, 88 : 265-356.

MHAISEN, F. T., AL- SALIM, N, K. and KHAMEES, N. R. (1986 recd. 1987) The Parasitic fauna of two Cyprinid and a mugulid fish from Mehaijeran Geek, Basrah (Iraq). *J. Biol. Sci. Res.*, 17 (3) : 63-74.

MITCHELL, L. G. (1967). *Myxidium macroheilin.* sp. (Cnidospora : Myxidium) from the large Scale sucker *Catostomus* macrocheilus Girad and a synopsis of the *Myxidium* of North American freshwater Vertebrates. *J. Protozool.*, 14 : 415-424.

MITCHELL, L. G. (1977). In : *Parasitic Protozoa*, Vol. IV (J. P. Krier ed.) Academic Press, Inc. New York, San Francisco, London.

MOSEVICH, T. N. (1965). Electron microscopic study of the structure of the contractile vacuole in the Ciliate *Ichthyopthirius multifieiis* (Fouquet). *Acta Protozool.*, 3 : 61-67.

MUKHERJEE, M. and HALDAR, D. P. (1981). Studies in the spore morphology of Myxobolus punctatus Ray Chaudhuri and Chakravarty, 1970 (Myxosporidia : Myxobolidea); Variations due to invasion in different organs of the fish *Ophicephalus punctatus* Bloch. *Arch. Protistenk.*, 124 : 29-35.

NAVILLE, A. (1930). Recher Ches sur la Sexualite Chez les Myxosporidies. *Arch. Protistenk.*, *Lxix* : 327-400.

NIGRELLI, R. F., POKORNY, K. S. and RUGGIERI, G. D. (1976) Notes on *Ichthyophthirius multifiliis*, a Ciliate parasitic on fresh water fishes with some remarks on possible physiological races and species. *Trans. Amer. Micros. Soc.*, 95 : 607-613.

NOBLE, E. R. (1943). Nuclear Cycles in the protozoan parasite *Myxidium gasterostei* n. sp. *J. Morphol.*, 73 : 281-295.

OTTO, G. R. and JAHN, T. L. 1943. Internal myxosporidian infections of some fishes of the Okoboj region. *Proc. Iowa Acad. Sci.*, 50 : 323-355.

PAPERNA, I., HARTLEY, A. M. and CROSS, R. H. M. (1987). Ultrastructural studies on the plasmodi of *Myxidium giardi* (Myxospore) and its attachment to the epithetlium of the urinary bladder. *Int. J. Parasitol.*, 17 (3) : 813-820.

PARISI, B. (1912). Primo contributo alla distribusions geografica dei missopporidi in Italia. *Att. Soc. Ital. Sc. Nat.*, 50 : 283-299.

QADRI, S. S. (1969). On a new myxosporidian *Henneguya Jubili* n. sp. from fresh water fish, *Notopterus notopterus* of Andhra Pradesh, India *Progress* in Protozoology, Abstract, p. 241.

RAY, H. N. (1933). Myxosporidia from India *Proc. Indian Sci. Congr.* Section IV Zoology p. 259.

RAY, H. N. (1933a). Preliminary observations on Myxosporidia from India. *Curr. Sci.*, 1 : 349-350.

RAY CHAUDHURI, S. and CHAKRAVARTY, M. M. (1970). Studies on Myxosporidia (Protozoa, Sporozoa) from the food fishes of Bengal, 1. Three new species From *Ophicephalus punctatus* Bloch. *Acta Protozool.*, 8 : 167-173.

RICE, V. J. and JAHN, T. L. (1943). Myxosporidian parasites from the gills of some fishes of the Okoboji region. *Proc. Iowa Acad. Sci.*, 50 : 313-321.

ROQUE, M. and PUYTORAC, P. DE. (1968). Infraciliature d'un nouvel Ophryoglenidae : *Ichthyophthirioides browni* n. g., n. sp. Protistologica, 3 (year 1967) : 465-473.

RUNNELLS, R. A., MONLUX, W. S. and MONLUX, A. W. (1976) *Principles of Veterinary Pathology*. Scientific Book Agency, Calcutta, pp. 1-958.

SARKAR, N. K. (1982). On three new Myxosporidian parasites (Myxudoa) of the Ophicaphalid fishes of west Bengal, India. *Acta Protozoologica*, 21 (3-4) : 239-244.

SARKAR, N. K. (1985). Some Myxosporida (Myxodoa:Myxospora) of anabantid fishes of West Bengal, India. *Acta Protozool.*, 24 (2) : 175-180.

SARKAR, N. K. (1985a). Some Coelozic Myxosporidans (Myxozoa: Myxosporea) from a fresh water teleost fish of river Padma. *Acta Protozoologica*, 24 (1) : 47-53.

SARKAR, N. K. (1986). A new Myxosporidan *Myxidium sciaenae* new species (Myxozoa : Myxididae) from the gall bladder of a marine teleost of West Bengal, India *Acta Protozool.*, 25 (4) : 477-479

SARKAR, N. K. and CHOUDHURY, S. R. (1986). The *Lohane llus beggalens* is sp. n. and *Myxidium mystusium* sp. n. (Myxozoa) : two new Myxosporidia from Indian fresh water teleost. *Acta Protozoologica*, 25 (3) : 359-362.

SARKAR, N. K., MAZUMDER, S. K. and PRAMANIK, A. (1985). Observations on 4 new species of Myxosporidia (Myxozoa) form Channid (ophicephalid) fishes of west Bengal, India. *Arch Protistenkd.*, 130 (3) : 289-296.

SHULMAN, S. S. (1966). The Myxosporidia of the fauna of USSR. *Akademiya Nauk USSR, Institut, Zoologii, Moscow* : 3-504.

SHULMAN, S. S. and STEIN, G. A. (1962). Phylum Protozoa. In : *Key* to *Parasites of Freshwater Fish of the U.S.S.R.* (I. E. Bykovskaya- Pavlovskaya et. Al. eds.) Izd. Akad. Nauk. SSR, Moscow- Leningrad. (Israel Program Sci. Transl. 1964, OTS, U.S. Deptt. Of Commerce.).

SIAU, Y. (1972). Mysoxporidies de *Synodontis ansorgii* Bouleng. Ann. et Mag : n. h. 1911 et de *Eleotns (Kribia) Kribensis* Boulenger 1964, Poissons des eaux saumatres de la *Bulletin Soc. Zool. Fr.*, 96 (4) : 563-570.

SLUKHAI, V. V. (1975). On the Syste matic position of some species of the genus Myxidium (Class : Cnidosporidia). *Problemy Parasct.*, 2 : 175-176

SRIVASTACA, C. B. (1975). Fish Pathological studies in India a Brief review. *Dr. B. S. Chauhan Comm.* Vol. pp. 349-359.

THELOHAN, P. (1892). Observations Sur les Myxosporidies et essai de Classification de Ces Organismes. *Bull. Soc. Philom.*, 4 : 165-178.

THELOHAN, P. (1895). Recherches Sur les Myxosporidies. *Bull. Sci. Fr. Belg.*, 26 : 100-394.

TRIPATHI, Y. R. (1955) Experimental infection of Indian major carps with *Ichthyophthirius multifiliis* Fouquet. *Curr. Sci.*, 24 : 236-237.

VAN DUIJN, JNRC. (1973). *Diseases of Fish.* 3rd ed. The Butterworth and Co., Publishers, Ltd., London. p. 372.

VENTURA, M. T. and PAPERNA, I. (1985). Histopathology of *Ichthyophthirius multifiliis* infection in fishes. *J. Fish. Biol.*, 27 (2) : 185-203.

YASUTAKE, W. T. and WOOD, E. M. (1957). Some Myxosporidia found in Pacific Northwest Salmonids. *J. Parasit.*, 43 : 633-642.

Key to Lettering of Text Figures

PC : Polar Capsule

PF : Polar Filament

S   : Sporoplasm

Sn : Sporoplasmic nucleus

Sv : Spore value

Vs : Value sture.

□□□

SECTION 2

WILDLIFE

13

URBAN IMPACTS ON THE WATER QUALITY AND THE MACROBENTHIC INVERTEBRATE FAUNA OF RIVER TAWI, J&K STATE

Raheela Mushtaq and Anil K. Verma

INTRODUCTION

Water is an essential condition of life on this planet. Water resources have been a dicisive factor in the growth and development of human civilizations throughout the history. In fact all the ancient civilizations were distinctly and predominantly hydraulic in nature and flourished and also perished in famous river valleys of the world due to the use and misuse of water resources.

Freshwater is a renewable resource by virtue of the hydrologic cycle, but for all practical purposes it is finite one. Freshwater comprises of springs, rivers, streams, ponds and lakes etc. All these habitats are biologically productive and useful to life as a support system and for regulation of atmospheric temperature. Freshwater habitats are biologically productive and support a web of life, which is not only diverse but also helps us to conceptualize the meaning and functioning of any ecosystem. For understanding of habitats like a lake or a pond as a demonstrative example, a lot has by now been investigated limnogically to a considerable extent. Essential food stuffs for meeting the needs of organisms inhabiting an aquatic ecosystem undergo continuous more or less definite cycles of changes indicating unity of a self contained institution, enjoying a considerable independence from the adjacent land.

Natural waters such as ponds, lakes and rivers form balanced ecosystems. They constitute an environment with set physical and chemical characteristics that sustain the aquatic communities and fluctuate in tune with the metabolic rhythms of the organisms. This well established metabolic rhythm of the organisms gets disturbed due to the variations in the present physical and chemical characteristics which is mainly caused by the water pollution.

River pollution is generally caused by the point sources and non-point sources and by far the most important of all the sources are the industries. Municipal waste is a source of water pollution next in importance only to the industrial wastes. From the very beginning of the civilization the only method in vogue for the disposal of sewage has been to dump it in nearby streams. For the assessment of effects of water pollution on living organisms, selection of a suitable system is required which may be used to elucidate the interplay of various complex factors.

The impact of the changes in the physical and chemical characteristics of aquatic ecosystems is clearly revealed by the inhabiting flora and fauna. The faunal components of an aquatic ecosystem include the zooplankton and the benthos. Of these the macrobenthic invertebrates are very good pollution indicators as they respond very well to the subtle changes of the water quality in an aquatic ecosystem and are thus also referred to as the efficient bioindicators. In order to assess the impact of the urban influences on the water quality of the river tawi, a comparative analysis of the macrobenthic invertebrate fauna residing in a selected stretch of the Tawi river was undertaken.

STUDY AREA

The main drainage of Jammu district is by the river Chenab. It is formed by the confluence of Chandra and Bhaga rivers rising from the opposite side of the Baralacha Pass at Tandi located in the upper Hmalayas in the Lahul and Spiti districts of Himachal Pradesh at an elevation of nearly 5200 m.s.*l.* This river has a torrential flow in Doda and Udhampur districts but meanders with slow currents in Jammu District. The river Chenab is joined by two important tributaries of which river Tawi is the longest and major tributary of river Chenab. It rises in middle Himalayas below the Dhar peak at Kalikund near Bhaderwah. After flowing through Chenani and Udhampur towns in Udhampur district, it enters Jammu district and in its upper catchment area it receives large number of other tributatries viz., Birhum khud, Ramnagar wali khud, Saloh nallah, Jhajjar nallah.

MATERIAL AND METHODS

For the present investigation two sampling stations SI and SII (Figs. 1-2) were selected at a stretch of 4 km along the longitudinal profile of the river tawi which exhibited variations in the ecological conditions and differential urban influences as well.

FIG. 1&2: Showing the collecting sites along the selected stretch of river Tawi at Jammu.

At SI, the substratum was comprised of the stones, pebbles and sand. The site was under lesser anthropogenic influences as compared to the SII which was under relatively greater anthropogenic influences like bathing of humans and animals, domestic sewage, waste from city crematorium and sand mining etc. (Fig. 3.).

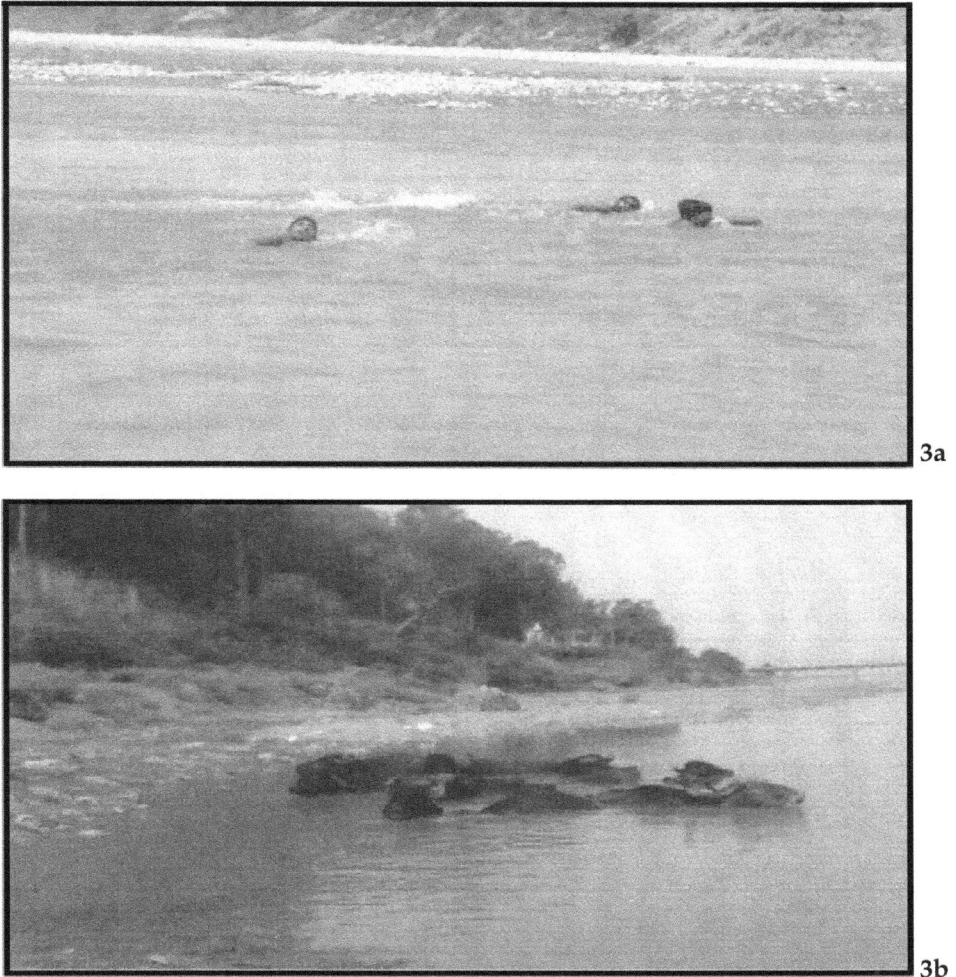

FIG. 3:   (a-b) showing extent of anthropogenic pressures on the stretch of river Tawi at Jammu.

The macrobenthic invertebrates were collected monthly using Ekmann's dredge and the collected samples were then sieved through sieve no. 40 having 256 meshes/ cm² (Edmondson and Winberg, 1971). After sieving, the material was packed in labeled polythene bags at the study site. The collected samples were washed in the laboratory and the organisms then sorted with clean water and preserved in 5% formalin for subsequent identification. Standard techniques were followed for the physicochemical analysis of the water (APHA, 1985) viz. temperature, depth, speed, pH, DO, FCO_2, CO_3^{2-}, HCO_3^-, Ca^{2+} Mg^{2+} and BOD during the present study.

OBSERVATIONS AND DISCUSSION

The macrobenthic invertebrate fauna of river Tawi recovered during the present study was represented by 3 major phyla viz. Arthropoda, Annelida and Mollusca. A total of 20 species were identified from both the stations of the stream of which 19 species were encountered at SII and only 9 species at SI. The composition of the species at the two sampling stations has been given in Table 1.

+ Present

- Absent

Table 1: Check list of the macrobenthic invertebrates encountered at the two stations of river Tawi

S. No.	Organisms	Station I	Station II
	Annelida		
1	Dero sps.	-	+
2	Stylaria sps.	-	+
3	Chaetogaster sps.	-	+
4	Lumbriculus sps.	-	+
5	Aeolosoma sps.	-	+
6	Tubifex tubifex	+	+
7	Eropobdella octobulata	-	+
	Arthropoda		
8	Hydropsyche sps.	+	+
9	Chauliodes sps.	+	-
10	Simulium sps.	-	+
11	Chironomous chironomous	+	+
12	Baetis rhodani	+	+
13	Caenis caenis	-	+
14	Cleone sps.	-	+
15	Deronectus sps.	-	+
16	Ophiogumphus	+	+
17	Macrobrachium kristensis	-	+
	Mollusca		
18	Lymneae accuminarta	+	+
19	Physa gyrina	+	+
20	Gyraulus convexiusculus	+	+

DISCUSSION

River Tawi is the principal drainage system of Jammu district. During its flow through the Jammu city, this river comes under numerous anthropogenic influences (Figs. 3a-b). The impacts of these stresses have resulted in the quantitative and qualitative variations of the macrobentic invertevrates population of the two stations. Qualitatively, out of the total 20 species, 19 species (95%) were found at S II and just 9 species (45%) were found at SI. Quantitatively a total of 595 ind/m² were recorded from both the stations out of which SI contributed 141 ind/m² & SII represent 454 ind/m² as put forth by Bhatt and Pandita (2006), the lower reaches of the stream are categorized to have low flow, temporary film of silt and clay together with organic detritus and reduced water currents. More or less similar conditions were noticed at SII of river Tawi during the investigation which provided a static and stable environment and it has proved to be very helpful for them to grow, reproduce and flourish. Such a view was also explained earlier by Kownacka and Kownacki (1965); Vasisht and Bhandal (1979); Sunder and Subla (1986); Dutta and Malhotra (1986) and Gopal and Sah (1996) from elsewhere.

Another notable observation made during the present investigation is the occurrence of the *Macrobrachium kristensis* at SII for a period of just two months of the year. This may be attributed to the increasing temperature and to the breeding season of this species. In addition to this slow water speed, organically rich bottom and thick vegetation cover at SII may be regarded as the suitable reason for the existence of these organisms at SII only. These findings are consistent with the findings of Chalotra (2002).

For the molluscan population, a peak was observed during the early summer months of which the quantitative abundance was noticed at SII as compared to SI which could be attributed to the presence of high concentration of Ca^{2+} HCO_3^- which is essential for the shell formation.

Thus relatively greater abundance of macrobenthic invertebrates were recorded, both quantitatively as well as quanlitatively, at SII as compared to SI. The river basin at SII, being under massive urban influences, receives large amounts of biologically active nutrients from the catchment area which is known to affect the distribution and abundance of benthic fauna(Marshall, 1971), and also the vegetation and amount of dissolved calcium. Apart from this soft substrata at SII could also be another important factor determining the abundance of organisms. Berg (1937) is also of the opinion that nature of bottom sediments influences the spread of the organisms.

It may thus be concluded that macrobenthic invertebrates prefer silty sandy and clayey organically rich bottom. The dependence of the macrobenthic invertebrates on such factors is very prominent because the substrate provides not only the habitat but also the food either directly or indirectly through the surface where food particles not only concentrate but also get protection from the predators. Such a view has also been maintained by a number of workers prominent among whom include, Egglishshaw and Mackay (1967); Learner *et al.* (1971); Hawkes (1979); Thakial (1997); Sharma (1999); Sharma (2002); Nelofer (2003) and Sawhney (2004).

ACKNOWLEDGEMENTS

We are grateful to the Principal G.D.C., Rajouri for providing necessary facilities at the Department of Zoology for compiling this work.

REFERENCES

APHA, (1985). Standard methods for the examination of water, 17th Ed. American Public Health Association.

Chalotra, R. (2002). Studies on early life history of freshwater prawn *Macrobrachium dayanum*. M. Phil Dissertation. University of Jammu.

Dutta, S. P. S. and Malhotra, Y. R. (1986). Seasonal variations in macrobenthic fauna of Gadigarh stream, Miran Sahib, Jammu. Ind. J. Ecol., 13 (1) : 138-145.

Edmondson, W. T. and Winberg, C. G. (1971). A manual on methods for assessment of secondary production in Fresh waters. IBP Hand book No. 17 : 358.

Egglishaw, H. J. (1968). The quantitative relationship between bottom fauna and plant detritus in streams of different calcium concentration. J. Appl. Ecol., 5 : 731-740.

Egglishaw, H. J. (1969). The distribution of benthic invertebrates on substrate in fast flowing streams. J. Anim. Ecol., 38 : 19-33.

Gopal, B. and Sah, M. (1993). Conservation and Management of rivers in India : Case study of the river Yamuna. Environmental Conservation, 20 (2) : 243-253.

Hawkes, H. A. (1979). Invertebrates as indicators of river water quality. In : Biological indicators of water quality edited by A. James and Lilian Evisan.

Kownacka, M. and Kownacki, A. (1965). The bottom fauna of the river Bialka and its Tatra tributaries, the Rybi Potok.

Learner, M. A., Williams, R. Horcap, M. and Hugher, B. D. (1971). A study of the macrofauna of the river Cyon, a polluted tributary of the river Traff (South Wales). Mackay, R. J. and Kalff, J. (1969). Seasonal variation in standing crop and species diversity of insect communities in a small Quebec stream. Ecology, 50(1) : 101-109. Fresh water Biol. J. : 339-376.

Mackay, R. J. and Kalff, J. (1969). Seasonal variation in standing crop and species diversity of insect communities in a small Quebec stream. Ecology, 50 (1): 101-109.

Nelofar, N. (2003). Limnology of a high altitude Sarkoot pond (Kishtwar) M. Phil Dissertation, University of Jammu, Jammu.

Sawhney, N. (2004). Limnological assessment of Ban-Ganga stream with reference to some consumers inhabiting the stream. M. Phil. Dissertation University of Jammu.

Sharma, A. (1999). Limnological studies of Ban-Ganga and distributional pattern of stream bottom fauna. Ph.D. thesis, University of Jammu, Jammu.

Sharma, S. P. (2002). Studies on the impact of anthropogenic influences on the ecology of Gharana Wetland, Jammu. Ph.D. thesis, University of Jammu, Jammu.

Sunder, S. and Subla, B.A. (1986). Macrobenthic fauna of a Himalayan river. Indian J. Ecol., 13(1) : 127-132.

Thakial, M. R. (1997). Studies on benthos in some habitats of Jammu. Ph. D thesis, University of Jammu, Jammu.

Vasisht, H. S. and Bhandal, R. S. (1979). Seasonal variation of benthic fauna in some North Indian lakes and Ponds. Ind. J. Ecol., 6 (2) : 33-37.

□□□

14

FEATHERS ARE RUFFLED BY WINDS
OF CHANGE IN KASHMIR
THE VALLEY OF BIRDS

B.L. Kaul

Sitting in the company of eminent physicians and naturalists Dr. Nasir Ahmed Shah and Dr. Girija Dhar in their beautiful garden at Kral Sangar, Brein Srinagar this summer I felt enchanted by the sweet resonance of the songs of Posh-i-nul* (Golden oriole) Kostur* (Tickle's thrush), Bill-bi-chur* (Bulbul), Kukil* (Dove) and the Fhamba-seer* (Fly catcher). When I expressed what I felt Syed Abdul Wahid, a retired senior civil servant and a friend who was also sitting with us reminded me of his school days when the legendary bird watcher of Kashmir Late Mr. Samsar Chand Koul used to take students of his school for bird watching to different gradens early in the morning. He had taught his students how to identify birds from their calls.

"masterji is still remembered by his students for influencing a whole generation of Kashmiris and for turning them into nature lovers and naturalists", said Dr. Nasir himself a student of Mr. Koul. It is indeed a tribute to his memory that, on "World Migratory Bird Day" organized by DR. C. M. Seth of WWF (J & K Chapter) in the month of May this year at Jammu, a section exhibiting photographs of rare and endangered migratory birds was dedicated to Mr. Samar Chand Koul.

Male Mallard

Bird watching and listening to their songs in gardens, meadows and forests can be a fascinating experience. At C. M. S. School Srinagar Rev. Canon C. E. Tyndall-Biscoe, the pioneer of modern education in Kashmir, used to lead parties of boys over mountains and through valleys and thus inculcated in them a love of nature. Thus trekking, bird watching, swimming, water sports besides other activities formed an essential part of education at C. M. S. (Now Tyndall-Biscoe memorial School) even after he had departed.

The State of Jammu and Kashmir has a rich and varied bird fauna. There are 262 species of birds reported from Kashmir valley, 225 from Ladakh and 183 from Jammu region. Unfortunately, however, recent studies have shown that there has been a decline in the populations of bird species. For example there are fewer Kites

---

*Local names

flying in the sky now than before. Vultures which were a common sight in the semi-urban and rural areas of the state are rarely found now thanks to the excessive use of Diclofenac by veterinarians to treat domestic animals for plain relief. By consuming the carcases of excessively drugged domestic animals the vultures got poisoned and killed. According to Mr. Wahid even common birds like house sparrows, mynas, pigeons and crows have become rare in Kashmir.

The reasons behind this phenomenon are not difficult to understand as they are purely man made. Earlier when houses were built some holes and spaces were left in the walls for birds like sparrows, mynas and pigeons to build their nests. Not so now. All modern buildings are fully plastered from outside and leave no spaces for birds to nest. I still remember that in my home every summer pairs of swallows used to come and make use of the nests built by them earlier after making necessary repairs with fresh mud and hay. The wooden ceilings of our rooms provided them an ideal location for nest building. We always left a window open for them to make their forays. Now RCC lintels have replaced wooden ceilings which are not suitable for nest building by swallows. I hope swallows still use the old houses for nest building.

The difficulty for birds to make their nests in fact started a few decades ago when cities and towns started expanding under the pressure of population explosion. It necessitated cutting down of trees which earlier provided not only cover but also safe sites for nest building. I still remember that in nineteen seventies a pair of kites used to lay eggs in nest atop an old popular tree in our neighbourhood at Srinagar. When the owner required additional rooms for his expanding family he fell down the poplar. Next year the same pair of kites finding the old poplar missing built a new nest from sticks on the G. I. sheet roof of a neighbouring house which luckily had a horizontally fixed wooden plank which provided the necessary support for the nest. The owner of the house had obsensibly fixed the plank for preventing fresh snow from sliding down and possibly hitting some passerby but it came handy to the poor kites for building their nest.

For many years a pair of crows used to lay their eggs in a nest made of sticks, wires etc in the angle of one of the supports of balcony of our old house and raise their nestlings year after year. However, one summer a nephew of mine out of curiosity to see the nestlings ventured near the nest. The ever vigilant parent crows become furious and raised hue and cry and started pecking at any one entering or leaving the house. This continued till the nestlings grew up and flew away. Crows are bold but most birds like sparrows, swallows and pigeons are timid and prefer to desert their nests leaving behind the nestlings to starve and die.

In the year 2008 after returning home from England late in the month of October I observed an unusual phenomenon in my Jammu house. A pair of house sparrows that had been breeding for several years from March to August in a nest they had built in our kitchen chimney was still busy in the process of raising a single hatchling even at that late time of the year. This is unusual. I kept on watching till the little sparrow grew fully and flew away in November. This obviously had something to do with climate change. That year the summer had prolonged as a result of the process of global warming and climate change perhaps making insects on which the young are fed available for a longer period and encouranging my winged friends to raise yet another offspring even as late as October-November.

I have cited the above common place examples to bring home the influence that human interference and activity can have on wildlife. Climate change is also a direct result of excessive use of natural resources such as fossil fuels. According to the latest reports from the British Trust for Ornithology (bird studies), a study has revealed that climate change affecting the patterns of nesting in birds. It has reported for example that climate change is hitting robins. With warmer winters and earlier springs, these much-loved garden birds are now laying eggs in mossy nests almost a week earlier than they were 40 years ago. It is reported that same is the case with chaffinches and great tits. The study has also revealed that due to climate the insect caterpillars on which birds feed their young ones are emerging earlier due to warmth. So it was feared that the beautiful synchronisation of bird hatching dates with the appearance of caterpillars, which has gone on for countless generations, would be wrecked and the young would starve. But the birds have adapted themselves to the change and in turn started nesting earlier.

However, according to another study also made in Great Britain all birds may not be adapting to climate change. For example as temperatures rise birds that flourish in warmer climates are beginning to come into Britain from further south and some birds such as woodlarks, hitherto found only in southern England are spreading further north. I myself witnessed a bird nest built atop a mobile phone tower on the shore of Loch Awe at Cruachan in preference to trees during a recent visit to Scotland. On making enquiry from the staff of the power plant located there I learnt that the pair of grey herons had been coming all the way from Africa for nesting for the last three years finding both safety and abundance of fish on which they feed themselves and their young ones. On the contrary whole populations of birds found in northern England and Scotland such as puffins, terns and razorbills have failed to rare young ones due to non-availability of food.

Unfortunately for us due to lack of interest shown by our young biologists in making studies on breeding behavioural patterns of bird populations, with some exceptions, we do not have much data available with us. I hope the Departement of Wildlife and the concerned University Departments promote studies on birds in Kashmir as well as Jammu and Ladakh.

In Kashmir summers as well winters witness the arrival of a great variety of birds. In summer the Golden oriole comes from as far as Transvaal in South Africa, the Paradise flycatcher form java and the Bee eater from cape colony. The winter migrants such as Mallards (neluj and thuj*), Wigeon ducks and Pintail ducks (sokha pachhin*) come from far off places like Kashgar, Tomsk and Tobolisk (Siberia). Studies on bird migration made by Bombay Natural History Society in which I too participated in seventees and eightees of the last century have shown that some of our winter migrants also come from wet lands of Bharatpur in Rajasthan and Harike in the Punjab. It is a pity that instead of shooting them with our cameras we shoot them with guns just to please our taste buds.

There is a need to revive interest in study of birds in our younger generation in Jammu and Kashmir. Ornithology is a field of study which has a wide scope and can help our young scientists in finding jobs world wide, and bring laurels to themselves and to the State.

Acknowledgement : Daily Greater Kashmir, Srinagar J&K State, India, 31 Oct., 2010.

---

*Local names

REFERENCES

Ali, Salim, (1996). *The Book of Indian Birds*. B. N. H. S. Mumbai.

Ara, Jemal, (1970). *Watching Birds*, National Book Trust, India.

Campbell, W. D. (1959). *Bird watching as a Hobby*, Stanley Paul, London.

James, Fisher, (1946). *Watching Birds.* Pelican Books.

Koul, S. C., (1957). *Birds of Kashmir*, Normal Press Publication, Srinagar.

Riplay, S.D., (1961). *A Synopsis of the Birds of India and Pakistan*, B. N. H. S. Mumbai.

Stuart, Smith, (1945). *How to Study Birds.* Collins.

❑❑❑

15

AVIAN DIVERSITY OF ERINGOLE SACRED GROVE IN THE WESTERN GHATS OF KERALA

E. A. Jayson

ABSTRACT

Status, occurrence and diversity of avifauna of Eringole Sacred Grove, Kerala were studied from August 2006 to July 2009. As the area was small in size the method of total count was applied to assess the avifaunaa. Monthly observations of 2 to 3 days were carried out during all the years. A total of 65 taxa of birds were recorded from the Sacred Grove. Higher species diversity was noticed in the month of Jaunary and lowest in July. Similarly diversity index was highest in January and lowest in July. All other diversity parameters showed similar trend. Breeding of Travancore spur fowl was recorded during the months of January to May. The occurrence of a population of this ground nesting fowl in an isolated sacred grove of 10 ha showed its ability to sustain within a limited space.

Keywords : *Eringole sacred grove, Galliformes, Travancore red spur fowl, Kerala, India.*

INTRODUCTION

Birds of Kerala have been well studied by many authors in the past . Studies on birds in Kerala mainly concentrated on preparing inventories from the protected areas and also from other areas like, reserve forest and wetlands. Apart from the inventories, detailed studies on many individual species were also carried out by the researchers of University of Calicut. Faunal components of Sacred Groves in Kerala were also reported earlier. No detailed report on the status of avifauna in the Eringole Sacred Grove is available. The objective of this study was to find out the diversity of avifauna in the Eringole Sacred Grove, Perumbavoor, Kerala. This will help in identifying the endemic or endangered avifauna of the sacred grove and also to conserve it.

Study Area

The Eringole sacred grove (10 ha) is situated in the middle of the Perumbavur Municipal town (10°10' and 10°42' N and 46°15' and 46°53' E) in Central Kerala.

Methods

Observations on birds were made on two days in each month. As the study area is small is size the method of total count was employed. The birds were identified up to species level and the location of each bird was marked in grids. Along with the species identity, the number of birds in each sighting, the activity of the bird and any other pertinent observations on the bird were also recorded. Travancore red spur fowl (Galloperdix spadicea) was observed from February 2007 to February 2008 and the breeding season and behaviour was recorded.

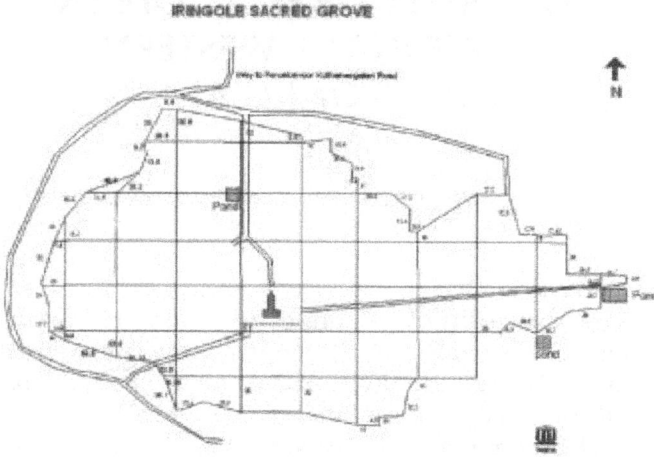

Fig. 1. Schematic sketch of the study area at Eringole sacred grove, Kerala.

Results

A total of 65 species of birds were recorded from Eringole sacred grove. The details of species recorded from the area are given in the Table 1. Among the birds recorded, seven are wetland species and another five are owls. Presence of a paddy field surrounding the grove and the presence of a pond inside the grove, supported the wetland species. No trans-continental migrant species were recorded. Over all diversity index parameters are given in Table 2. Shannon diversity index is 3.13. On an average, twenty five individual birds were recorded during the observation. Limited observations on, feeding revealed that most of the bird species were dependent on the trees for their food, followed by the pond in the premises and also the surrounding paddy fields. The total number of birds recorded in different months is given Fig. 2. Highest numbers of birds were seen in the month of July.

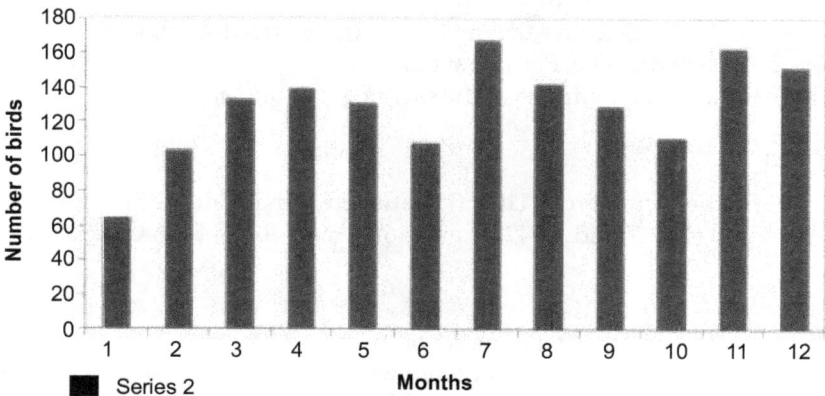

Fig. 2. Total number of birds recorded in each month

Table 1. Species of birds recorded from the Eringole sacred groves.

No.	Order	Family	Genus	Scientific Name	Common Name	IUCN status
1	Passeriformes	Corvidae	Dicrurus	*Dircrurus aeneus*	Bronzed drongo	LC
2	Passeriformes	Corvidae	Dicrurus	*Dicrurus macrocercus*	Black drongo	LC
3	Passeriformes	Corvidae	Dicrurus	*Dicrurus paradiseus*	Greater racket-tailed drongo	LC
4	Passeriformes	Corvidae	Dendrocitta	*Dendrocitta vagabunda*	Indian Tree pie	LC
5	Passeriformes	Corvidae	Corvus	*Corvus splendens*	House crow	LC
6	Passeriformes	Corvidae	Corvus	*Corvus macrorhynchos*	Jungle crow	LC
7	Passeriformes	Corvidae	Oriolus	*Oriolus xanthornus*	Blackheaded Oriole	LC
8	Passeriformes	Corvidae	Oriolus	*Oriolus oriolus*	Golden oriole	LC
9	Passeriformes	Corvidae	Pericrocotus	*Pericrocotus flammeus*	Scarlet minivet	LC
10	Passeriformes	Corvidae	Terpsiphone	*Terpsiphone paradisi*	Asian Paradise flycatcher	LC
11	Passeriformes	Corvidae	Aegithina	*Aegithina tiphia*	Common iora	LC
12	Passeriformes	Corvidae	Tephrodornis	*Tephrodornis pondiceranus*	Common wood Shrike	LC
13	Passeriformes	Irenidae	Chloropsis	*Chloropsis cochinchnensis*	Goldmentled chloropsis	LC
14	Passeriformes	Irenidae	Irena	*Irena puella*	Asian Fairy blue-bird	LC
15	Passeriformes	Sylviidae	Turdoides	*Turdoides affinis*	White-headed babbler	LC
16	Passeriformes	Sylviidae	Turdoides	*Turdoides striatus*	Jungle babbler	LC
17	Passeriformes	Sylviidae	Phylloscopus	*Phylloscopus trochiloides*	greenish leaf warbler	LC
18	Passeriformes	Sylviidae	Acroephalus	*Acroephalus dumetorum*	Blyth's reed warbler	LC
19	Passeriformes	Sylviidae	Orthotomus	*Orthotomus sutorius*	Common Tailor bird	LC
20	Passeriformes	Muscicapidae	Copsychus	*Copsychus saularis*	Oriental Magpie-Robin	LC
21	Passeriformes	Muscicapidae	Zoothera	*Zoothera citrina*	White-throated thrush	LC
22	Passeriformes	Muscicapidae	Muscicapa	*Muscicapa daurica*	Asian Brown flycatcher	LC

Contd...

Table 1. Contd...

No.	Order	Family	Genus	Scientific Name	Common Name	IUCN status
23	Passeriformes	Nectariniidae	Nectarinia	*Nectarinia asiaticus*	Purple Sunbird	LC
24	Passeriformes	Nectariniidae	Nectarinia	*Nectarinia zeylonica*	Purple rumped sunbird	LC
25	Passeriformes	Pycnonotidae	Pycnonotus	*Pycnonotus jocosus*	Red-whiskered bulbul	LC
26	Passeriformes	Pycnonotidae	Pycnonotus	*Pycnonotus cafer*	Redvented bulbul	LC
27	Passeriformes	Sturnidae	Acridotheres	*Acridotheres tristis*	Common myna	LC
28	Passeriformes	Sturnidae	Acridotheres	*Acridotheres fuscus*	Jungle myna	LC
29	Passeriformes	Pittidae	Pitta	*Pitta brachyura*	Indian Pitta	LC
30	Passeriformes	Passeridae	Dendronanthus	*Dendronanthus indicus*	Forest wagtail	LC
31	Passeriformes	Passeridae	Motacilla	*Motacilla cinerea*	Grey wagtail	LC
32	Passeriformes	Passeridae	Lonchura	*Lonchura malacca*	Black-headed munia	LC
33	Passeriformes	Passeridae	Passer	*Passer domestica*	House Sparrow	LC
34	Passeriformes	Strigidae	Strix	*Strix ocellata*	Mottled wood-owl	LC
35	Passeriformes	Strigidae	Glaucidium	*Glaucidium radiatum*	Barred jungle owlet	LC
36	Passeriformes	Strigidae	Ninox	*Ninox scutulata*	Brown hawk-owl	LC
37	Passeriformes	Strigidae	Otus	*Otus bakkamoena*	Collared scops owl	LC
38	Passeriformes	Tytonidae	Tyto	*Tyto alba*	Barn owl	LC
39	Passeriformes	Centropodidae		*Centropus sinensis*	Greater coucal	LC
40	Passeriformes	Cuculidae	Eudynamys	*Eudynamys scolopacea*	Asian koel	LC
41	Passeriformes	Cuculidae	Cuculus	*Cuculus micropterus*	Indian cuckoo	LC
42	Passeriformes	Columbidae	Streptopelia	*Streptopelia chinensis*	Spotted dove	LC
43	Passeriformes	Columbidae	Chalcophaps	*Chalcophaps indica*	Emerald dove	LC
44	Passeriformes	Columbidae	Treron	*Treron phoenicoptera*	Yellow-footed Green pigeon	LC
45	Passeriformes	Columbidae	Columba	*Columba livia*	Blue rock pigeon	LC
46	Passeriformes	Ardeidae	Bubulcus	*Bubulcus ibis*	Cattle egret	LC
47	Passeriformes	Ardeidae	Egretta	*Egretta garzetta*	Little egret	LC
48	Passeriformes	Accipitridae	Haliastur	*Haliastur indus*	Brahminy kite	LC
49	Passeriformes	Phalacrocoracidae	Phalacrocorax	*Phalacrocorax niger*	Little cormorant	LC

Contd...

Table 1. Contd...

No.	Order	Family	Genus	Scientific Name	Common Name	IUCN status
50	Passeriformes	Halcyonidae	Halcyon	*Halcyon smynesis*	White-breasted kingfisher	LC
51	Passeriformes	Halcyonidae	Halcyon	*Halcyon capensis*	Brownheaded storkbill	LC
52	Passeriformes	Cerylidae	Ceryle	*Ceryle rudis*	Pied kingfisher	LC
53	Passeriformes	Alcedinidae	Alcedo	*Alcedo athis*	Small blue kingfisher	LC
54	Passeriformes	Meropidae	Merops	*Merops orientalis*	Small green bee-eater	LC
55	Passeriformes	Coraciidae	Coracias	*Coracias benghalensis*	Blue jay	LC
56	Passeriformes	Phasianidae	Perdicula	*Perdicula erythrorhyncha*	Painted Bush-quail	LC
57	Passeriformes	Phasianidae	Galloperdix	*Galloperdix spadicea*	Red spur fowl	LC
58	Passeriformes	Megalainmidae	Megalaima	*Megalaima viridis*	White-cheeked barbet	LC
59	Passeriformes	Picidae	Dinopium	*Dinopium benghalense*	Lesser Golden-backed Wood pecker	LC
60	Passeriformes	Picidae	Dinopium	*Dinopium javanense*	Indian goldenbacked threetoed woodpecker	LC
61	Passeriformes	Psittacidae	Psittacula	*Psittacula krameri*	Roseringed parakeet	LC
62	Passeriformes	Psittacidae	Psittacula	*Psittacula cyanocephala*	Blossomheaded parakeet	LC
63	Passeriformes	Psittacidae	Loriculus	*Loriculus vernalis*	Lorikeet	LC
64	Passeriformes	Upupidae	Upupa	*Upupa epops*	Hoopoe	LC
65	Passeriformes	Apodidae	Apus	*Apus melba*	Alpine swift	LC

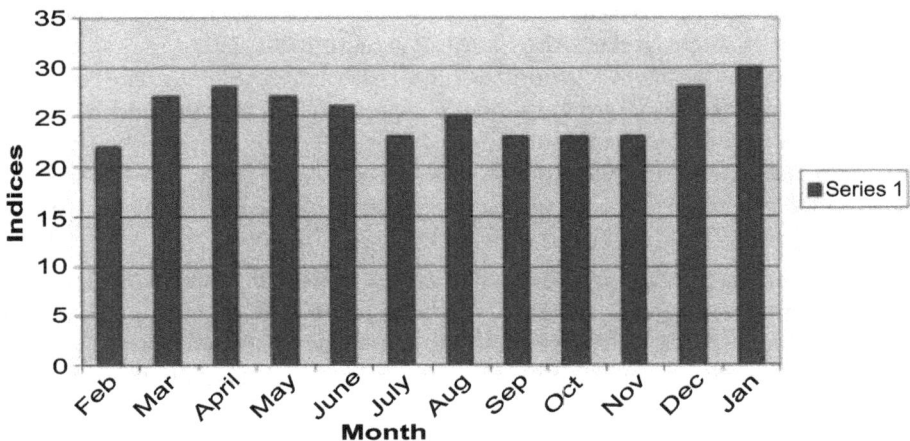

Fig. 3. Distribution of number of species in different months.

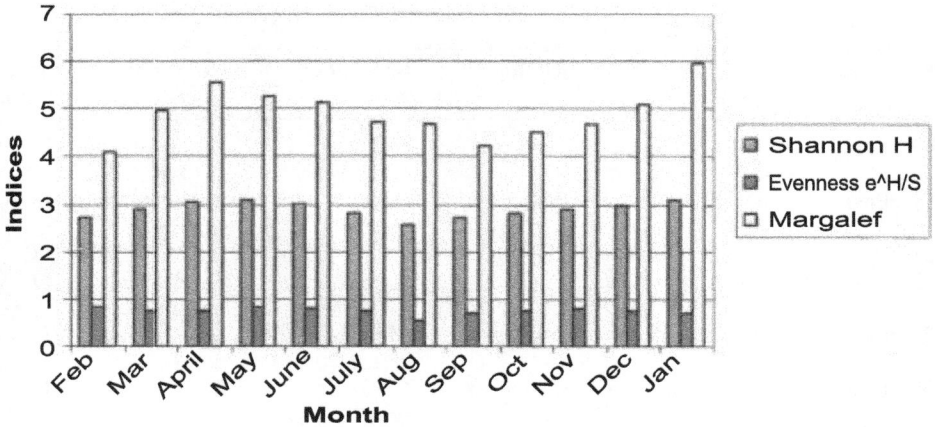

Fig. 4.　Diversity indices during different months.

Table 2. Diversity parameters of birds in the sacred grove

Total birds sighted	Dominance Index (D)	Shannon diversity index (H)	Evenness index	Equitabilities (J)
1536	0.05692	3.13	0.63	0.8737

Breeding of Red Spur Fowl

Red spur fowl is one of the three species of the genus Galloperdix which is distributed throughout the Indian subcontinent and endemic to the country. The species includes three subspecies in India (G. s. spadicea, G. s. stewarti and G. s. caurina). Travancore red spur fowl G. s. stewarti is a resident bird of Kerala found commonly in the forest areas excluding the Wayanad Wildlife Sanctuary. It prefers moist deciduous forests dominated by Lantana scrubs and undergrowth in the bamboo plantations (Ali and Ripley, 1983). Information on the ecological and behavioural aspects of this elusive and shy birds is scarce even though the species is reported from throughout the State (Easa and Jayson, 2004). Three eggs were found in a nest, which was built on the ground near the base of a tree (Plate 1). The occurrence of the bird in a sacred grove surrounded by houses and roads in the middle of a city is worth mentioning. Nearest natural forest was 18 km away from the site as the crow flies (Thattekkad Bird Sanctuary) and there is also no continuity of forest. The nesting site has a canopy height of 8 canopy cover of 50%. Distance to the nearest road was 75 m and distance to the water body was 20 m (Plate 2).

It is observed that, the species breeds during April-May months and January-February months in the year. It is also observed that it breeds on the ground, with out any hiding (have seen sitting over the eggs in an open place with out any hide or canopy cover and easily seen to others or beneath the trees. Usually there are 2 to 3 eggs, looking like local hen eggs and have almost same size. It is observed that it does not make a nest like others. It simply lays eggs among litters and sits over it and does not move until humans disturb it with a stick or other things.

Plate 1. Travancore red spur fowl at Eringole sacred grove

Plate 2. Eggs of travancore red spur fowl at Eringole sacred grove

Table 3. Breeding season of Travancore red spur fowl in the sacred grove

Months	Number breeding sites found
January-2007	2
February-2007	2
March-2007	2
April-2007	1
May-2007	1
June-2007	-
July-2007	-
August-2007	-
September-2007	-
October-2007	-
November-2007	-
December-2007	-
January-2008	4
February-2008	4

As per the reports, the subspecies Travancore red spur fowl usually nest throughout the year except during the monsoon months from June to August (Ali and Ripley, 1983). Both the observations support the above hypothesis. Regarding the nest site, Travancore red spur fowl is a well-known ground nesting species and known to build nest on ground in dense bamboo or scrub jungle, some times sparsely lined with leaves and grasses (Ali and Ripley, 1983).

Discussion

As recorded from other tropical forests, the highest species richness and number of birds was recorded during the month of January and December. This is mainly due to the influx of migrants to the area. Diversity indices also showed higher values during these months. The lowest species richness and diversity indices were recorded during the month of south west monsoon i.e., June and July. All the species are in Least Concern (LC) status of IUCN. The occurrence of a population of the ground nesting fowl, Red spur fowl in an isolated forest patch (Sacred grove) of 10 ha extend showed the birds ability to sustain within this limited space. Proper protection of the area is needed for the preservation of birds species in the area.

ACKNOWLEDGEMENTS

The author is grateful to Dr. K. V. Sankaran, Director, KFRI and Dr. C. Renuka, Programme Coordinator for the facilities and encouragement. Many thanks are also due to the authorities of Eringole Sacred Grove, who gave permission for the study. The author is thankful to the two Research Fellows who worked in the project. The Ministry of Environment and Forests, Government of India provided financial assistance for the study.

REFERENCES

Ali, S. and Ripley, D. (1983). Hand Book of the birds of India and Pakistan. Oxford University Press, Oxford, p. 737.

Buceros-Envis news letter: Avian Ecology and Inland Wetlands- Vol. 6, No. 1(2001)-Standardized common and Scientific Names of the Birds of the Indian Sub-continent.

Cody, Martin L. (1971). Ecological aspects of reproduction. In. Avian Biology Vol I (Eds). D. S. Farmer and J. R. King 4562-503.

Easa, P. S. and Jayson, E. A. (2004). Biodiversity documentation for Kerala. Part II: Birds. Kerala Forest Research Institute, Peechi. Hand Book No. 17, p. 51.

Horn, H. S. (1968). The adaptive significance of colonical nesting in the Brewer black bird (*Euphagus cyanocephalus*). Ecology. 49, 682694.

Jayson, E. A. (1998). Studies on Man- wildlife conflict in Peppara Wildlife Sanctuary and adjacent areas. Kerala Forest Research Institute, Research Report No. 140. p. 71.

□□□

16

OBSERVATIONS ON FEEDING AND BREEDING BIOLOGY OF THE BROWN ROCK CHAT *CERCOMELA FUSCA* (BLYTH) IN JAMMU (J&K STATE) INDIA

B.L. Kaul

ABSTRACT

The Brown Rock Chat also called Indian chat is a common resident bird of the plains of Jammu region of Jammu and Kashmir State. The present study deals with the feeding and breeding biology of the bird. It feeds mainly on insects including ants, spiders and also forages on selected food left overs of the humans. It breeds in human habitations and also in rock crevices from February to August.

Keywords: *Habits, call, eggs, hatching, nestling, fledgling and juvenile.*

INTRODUCTION

The Indian State of Jammu Kashmir has a rich and varied avifauna. There are 262 species of birds reported from Kashmir Valley, 225 species from Ladakh and 183 species from Jammu region. Unfortunately, however, recent studies have shown that there has been a decline in the population of many birds species due to anthropogenic reasons in all the three regions of the state. The present study on the feeding and breeding biology of the Brown Rock Chat, a common resident bird of Jammu, Samba and Kathua districts was undertaken in Jammu district (32°-50° & 33°-30 N and 74°-24 & 75° -18 E) to ascertain whether this bird is facing any threat to its existence. The Jammu district covers an area of 3097 sq. km and lies at a height of 340-410 m from mean sea level (MSL).

The Jammu region is spread over the middle Himalayas (Pir Panjal and upper Chenab region, the Shivalik region and the outer plains). The outer plains include the parts of Kathua, Samba and Jammu districts. The outer plains and the outer hills are grouped into tropical and subtropical climate regimes and the chief towns are Jammu, Samba, Hira Nagar, Kathua, Billawar, Basohli and Akhnoor. These towns experience intense tropical heat in summer. Most of the rainfall comes from July to September. The annual rainfall is about 150 mm. The winter season in mild and extends from December to February.

OBSERVATIONS

Distribution

The bird under study namely the Brown Rock Chat is distributed in North-West, Central and Eastern India and in South up to Narmada River (Khacher, Luvkumar

2000). It also extends slightly into Nepal and Northern Pakistan where it is restricted to east of the Chenab river. Although largely resident some populations make movements in response to weather. In the foot hills of Himalayas it reportedly moves higher up in summer; appears in Dehradun in spring and leaves for the plains before the onset of winter (Ali and Ripley 1998).

Habits

The Brown Rock chat keeps to rocky hills, rivers, ruins and in the compounds of old residential houses. It is seen singly or in pairs. The bird is usually tame and is not afraid of human presence. In fact it prefers to stay within populated regions.

Description and Classification

The Brown Rock chat (Fig. 1) is larger than the somewhat similar looking female Indian Robin but lacks the reddish vent and is about 17 cm long. It is uniformly rufous brown with the wings and tail of a slightly darker shade. The brown on the undersides grades into a dark grey brown vent (Rasmussen and Anderton 2005). The beak is black slender and slightly curved up at the tip. The second primary is the largest and tail is rounded. In flight it resembles a female Blue Rock Thrush and Redstarts. The sexes are indistinguishable in the field. It is characterized by habit of slowly raising its tail, slightly fanning it and bobbing its heed. It belongs to the subfamily Saxicolnae of the family Musciapidae of the order Passeriformes. Indian chat is the only species of Cercomela found outside Africa and its placement has been questioned in the past (Oates 1890) and a recent study of molecular phylogeny has suggested its placement in the genus Oenanthe (Outlaw et al 2010)

Fig. 1.   An adult Brown Rock Chat

Call

The Brown Rock chat is a wonderful songster. In Jammu it is the earliest bird to sing. A bird regularly visits the roof of the author's house at Sarwal Jammu as early as 3.30 a.m. and sings in melodious notes occasionally mimicking other birds like thrushes. It has a wide repertoire of calls (Sethi and Bhatt 2008). The usual call is a short whistling Chee delivered with a rapid bob and stretch. The alarm call is a harsh check-check. The song is sweet thrush-like with a number of notes. It is a very good mimic and imitates songs of other birds like yellow-eyed Babbler, Grey Cuckoo Shrike and Tickel's Blue Flycatcher.

Feeding

It feeds mainly on insects picked off the ground (Donahue, 1962). The author found them feeding on ants, flies and spiders as well. They also rummage human food left overs from streets to pick off bits of biscuits, potato chips and broken egg shells (Fig. 2). It has been reported that they also feed late and forage on insects attracted street by lights (Singh 2009).

Fig. 2. An adult bird rummaging human food left overs. Pieces of broken poultry egg shell can be seen

Breeding

The breeding season starts from February and extends up to August. The nest is a rough cup of rootlets, grass, hair, feathers, pieces of polythene etc. placed in a rock cleft, a hollow in a wall of a deserted building, outside a window or even inside a hollow in a wall of an occupied house. In one case the author observed a nest built between the top rung of a moveable iron ladder and the wall supporting it. The nest is generally at a low level and not hidden from sight and is easily detectable. This makes the eggs and chicks vulnerable. These birds prefer to build a nest on a foundation of pebbles. In one case the author observed a nest built on small pieces of stones and pebbles lying outside window of a house under renovation (Fig. 4). The nests are built and guarded against intrudes by both the parents. The usual clutch is 3 to 4 pale blue eggs with rusty specks and spots (Fig. 3) the female lays one egg per day. When laid only the female in incubates the eggs (Fig. 4) while the male acts as

Fig. 3.   Nest of Brown Rock Chat with eggs

Fig. 4.   A Female bird hatching eggs

a watchman. The eggs start hatching after 14-15 days. The hatchling is reddish with a yellow beak (Fig. 5). It is blind. The hatchlings are fed by both parents on insects and spiders. The nestlings and fledglings have black plumage (Figs. 6 and 7). The fledglings start to move out of the nest for short durations and then return. This is a crucial period for them as they can fall prey to prowling cats and other predators. The black

Fig. 5.   Emergence of the first hatchling

Fig. 6.   An early nestling in the nest

colored juveniles with shades of brown (Fig. 8) leave the nest after about 20 days of hatching. The pair raises two or three broods per season. The old nest may be refurbished for reuse or deserted if success at incubation and hatching has been poor.

Fig. 7.   A Fledgling chick outside the nest after first flight

Fig. 8.   A juvenile (blackish brown)

Discussion and Conclusion

The present observations on feeding and breeding of Brown Rock chat have brought forth some very interesting facts. With regard to their feeding behaviour it was a surprise to the author to see them seeking ant holes and feed on ants which are normally avoided by most insectivorous birds because of the offensive formic acid present in their bodies. It is also interesting to watch them look for spiders in their cobwebs inside crevices in walls and in old deserted buildings. Equally interesting is their habit of foraging human food wastes for pieces of biscuits, potato chips, pieces of broken egg shells and even sweet meats to feed themselves. They feed their chicks mainly on insects caught on ground.

As rightly pointed out by Salim Ali (1996) "We have still a great deal to learn about the breeding biology of even some of our common birds. Egg collection alone is not enough". The celebrated ornithologist also felt that breeding biology involved studies like the share of sexes in nest-building biology on incubation, care of the young, nature and quality of food fed each day and behaviour of the parents and young. There is no doubt that since he wrote those lines more than thirty years ago, much ground has been covered in our knowledge of breeding biology of birds, thanks to the efforts of many ornithologists in India and elsewhere. We now know much more about reproductive biology of birds than before but a lot more needs to be known.

Breeding in birds is an interesting subject. Most birds are very secretive about making their nests, laying eggs, incubating and feeding their young. It is not easy to watch their activities and you need a discerning and curious eye. Bird watchers invest a lot in terms of their patience, labour and time to learn more and more about lives of birds.

Both the sexes take part in nest building inside a hollow in a wall, a rock cleft or a space outside a window of a kitchen or bathroom. The nests which are built at low height are made of roots, grasses and other materials available in the surroundings. Some soft materials including down feathers and pieces of polyethylene are also used in nest making. The female lays one egg per day and the job is over in three to four days. Then she singly starts incubating while the male watches from a nearby place to ward off intruders like palm squirrels. The first hatchling emerges after 14 or 15 days followed by others with a gap of one day. The process of hatching is over in 3 to 4 days. Both the parents start feeding hatchlings by turns at regular intervals. Within 15 days or so naked red coloured hatchlings turn into black colored fledglings which start moving out of the nest only to return after a while. Soon they start taking small flights. The juveniles look like their parents but for their blackish brown plumage. Soon their plumage turns brown and juveniles (Fig. 8) look like parents (Fig. 9).

The strategy of building an easily visible nest at a low height goes to the disadvantage of the Brown Rock Chat. In one instance the author found that in a nest built outside a window there were initially four eggs. One egg was stolen by somebody by the time the female started incubating. Ultimately only one egg hatched and two failed to hatch. The single hatchling, however, was fed by both the parents and it changed into a fledgling within two weeks. In another instance the author found a nest inside a hole in one of the walls of a kingdergarten class of a school, hardly 5 feet above the ground, in a densely populated area of Jammu city. The kids observed the nest (which had 3 eggs) and asked their teacher to let them touch the eggs. The concerned teacher told the kids that bird nests should not be disturbed and lifted them up one by one to show the eggs. As the eggs hatched and the hatchlings started making noises to demand food from the parents the kids watched with excitement. They watched the whole process of incubating feeding, change of hatchlings to nestlings and then to fledglings. They received the first lessons of bird watching in their own class room! It was fun for them to watch bird reproduction and at the same time they had learnt to respect wildlife. Had the same nest been in a class room of older children there was every probability of the nest being disturbed and eggs stolen.

Fig. 9.   The juvenile turns into an adult bird. Note the change of colour to dark brown

Contrary to the habit of building nests at low heights by the Brown Rock Chat the common house sparrow (Passer domesticus) is more clever and builds its nests deep inside holes in the walls, using even kitchen chimneys and every conceivable place not visible to the predators. Their success rate of hatching is also better as compared to the Brown Rock chat. Two to three clutches per season, however, compensate the Brown Rock Chat to maintain the level of their population.

The present studies show that the populations of the Brown Rock Chat are stable in Jammu region of the Indian State of Jammu and Kashmir. They are not threatened and their conservation status is of least concern. However, public awareness about need of conservation of birds needs to be created in view of their diminishing populations.

REFERENCES

Ali Salim (1996). The Book of Indian Birds (Revised Edition) Bombay Natural History Society , p. 281.

Ali Salim; Sidney Dillon Ripley (1998). Hand book of the birds of India and Pakistan. Volume 9 (2 Edition) New Delhi; Oxford University Press, p. 21-22.

Baker, ECS (1924). The Fauna of British India, including Ceylon and Burma. Birds. Volume 2 (2 ed) London, Taylor and Francis.

Birdlife International (2009). Cercomela fusca. International union for conservation of Nature. Redlist of threatened species. http://www.iucnredlist.org/apps/redlist/details/ 147559. 18 June 2011.

Deignan HG, RA Paynter Jr. & SD Ripley (1964). Check list of birds of the world. Volume 10 Cambridge, Massachusetls: Human Comparative Zoology.

Donahue, J. P (1962). "Field identification of the Brown chat". Newsletter of Bird Watchers 2(10): 5-6.

En. Wikipedia.org/Brown_Rock_chat (2012).

Khacher, Lavkumar (2000). "BrownRockChat Ceromela fusca: extension of range into Gujrat'. News letter for Birds watchers 40(3):41.

Koelz, W. (1939). Proc. Biol Soc. 52:66.

Mathews, W. H. (1919). "Nesting habits of the Brown Rock Chat Ceromela fusca J. Bombay Nat. Hist. Society 26(3): 843-844).

Oates, EW (1890). Birds In India. London: Taylor and Francis, pp. 79-80.

Outlaw, RK; G. Voelkar and RCK Ceromela (Muscicapridae) and its relation to Oenanthe reveals extensive polyphyly among chats distributed in Africa, India and Paleearctic". Moleculer Phylogenetics and Evolution 55(1) 284-292.

Rasmussen PC and JC Anderton (2005). Birds of South Asia. The Ripley Guide Volume 2. Smithsoian Institution and Lynx editions, p. 408.

Sethi, V.K., Bhatt, D (2008). "Call repertoire of an edemic avain species, the Indian chat Cercomela fusca". Current science 94(9): 1173-79.

Singh, P. (2009). "Unusal nocturnal feeding by Brown Rock Chat Cercomela fusca (Passeirformes: Muscicapidae) in Bikaner Rajesthan, India" Journal of threatened Taxa 1(4):25.

White , L. S. (1919). Nesting habits of the Brown Rock Chat Ceromela fusca J. Bombay Nat. Hist. Soc. 26(2): 667-668.

❑❑❑

17

STATUS AND CONSERVATION OF ENDANGERED BLACKBUCK, ANTILOPE CERVICAPRA OF ODISHA, INDIA

Sudhakar Kar

The Indian Blackbuck (*Antiope cervicapra*), is one of the three species of antelopes found in Orissa. Is is locally known as **'Krushnasarmruga', 'Bail harina' and 'Kala bahutia'** The other two antelopes are Nilgai (*Boselaphus tragocamelus*) and the Chowsingha (*Tetracerus quadricornis*). All the three are even-towed (Artiodactyla) Bovids. Blackbuck is considered to be the second fastest animal in the world next to Cheetah. There is a fast decline in the population of Blackbucks throughout the country due to poaching and habitat loss. In the recent past, this endemic animal was quite numerous and commonly seen. Subsequently, within a short span of time it has suffered a drastic reduction in numbers. Blackbuck is included in the Schedule-1 of Wildlife (Proctection) Act, 1972 and is designated as Vulnerable as per Red Data Book (1994). It is one of the most popular exhibits in the zoos of the country and elsewhere.

Fig. 1. An adult male Blackbuck with prominent spiral horns

DISTRIBUTION AND STATUS

In India the species is widespread in Rajasthan, Gujarat, Madhya Pradesh, Tamil Nadu and other areas throughout peninsular India. In 1982, the estimated population in India was between 22,500 to 24,500. According to 1993 estimation, the population of Blackbuck in the country was about 10,000 and was stable.

Past Distribution in Odisha

This species was occurring in Balesore and Puri districts and very scarcely in Bolangir and Kalahandi districts and also in coastal sand dunes of Bhitarkanika and Kujang area. Upto the 1960s, the Blackbuck number was reported to be 1200-1300.

Present Distribution in Odisha

It is now confined to Balukhand-Konark Sanctuary in Puri district (small number); Balipadar-Bhetnoi and adjacent areas in Ganjam district.

In odisha the estimated population of Blackbuck is about 1100 to 1200. The population is on increasing trend in Ganjam district.

Fig. 2.   A herd of Blackbucks in the paddy fields at Balipadar

The **Balipadar-Bhetnoi** Blackbuck is seen in about 80 villages of Buguda, Aska and Kodala Forest Ranges in Ganjam District. The Blackbucks of Balipadar-Bhetnoi area are proctected religiously by the local communities. The belief that the presence of Blackbuck in the paddy fields brings prosperity to the village has contributed greatly to the protection of this species. The villagers do not kill the animal even if it strays into the fields and grazes their crop. Such protection has been approved by the local people for several generations. As the story goes more than a century ago, there was once a long spell of drought in this locality. During this period, a small group of Blackbuck appeared from somewhere in the area for the first time, and then there was rain and the drought spell was broken. Since then people have been rigidly protecting these animals as they feel that their fate is linked with the Blackbuck. During 1918, a Britisher known as "Green saheb" and the 'Sardar'of the locality Sri Madeshi Chandramani Dora took the initiative for protection of this species and published a notification in the Oriya news paper "Prajamitra" prohibiting killing of the Blackbuck. Recent census has put their number in this area at 1101, and there is a healthy growth in the numbers over the past decade.

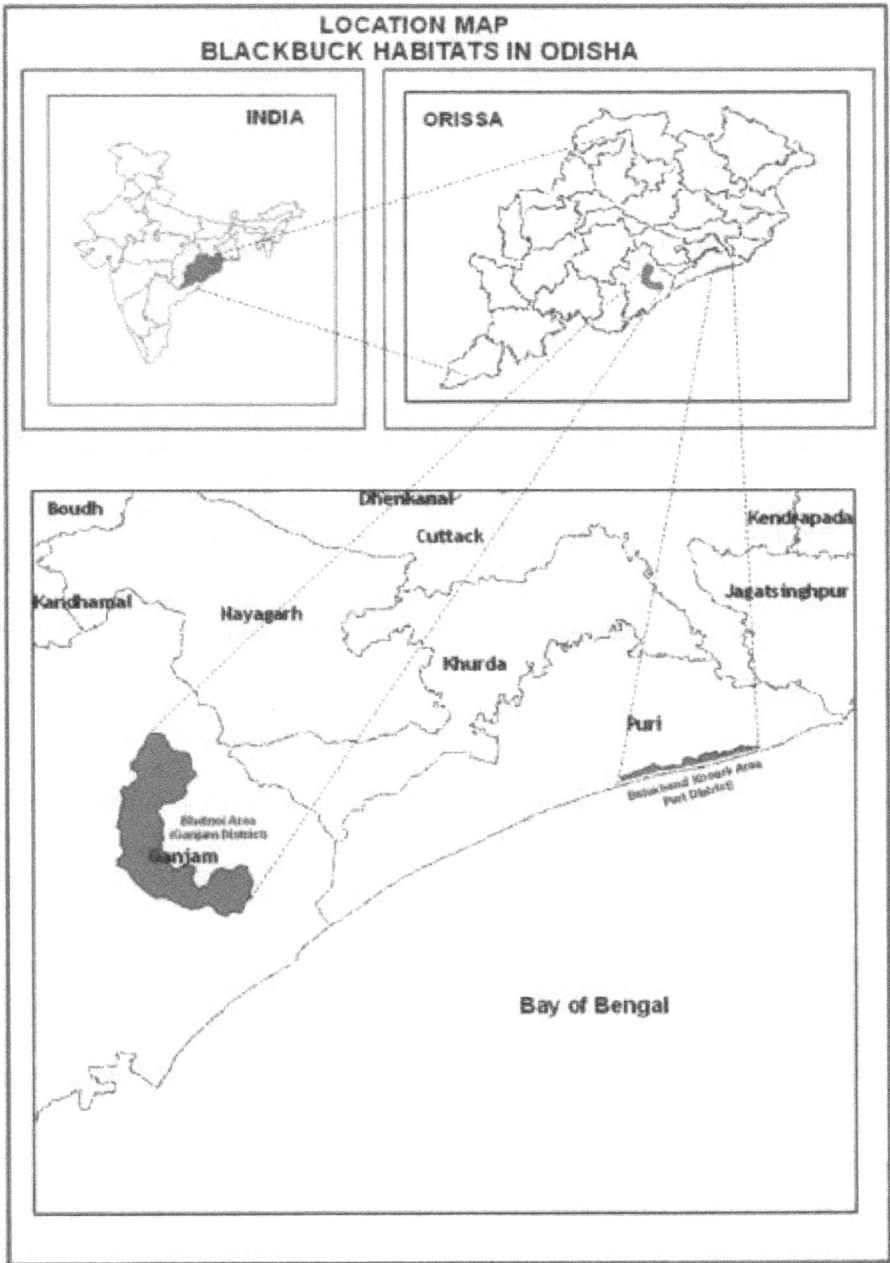

LOCATION MAP
BLACKBUCK HABITATS IN ODISHA

INDIA

ORISSA

Boudh

Dhenkanal

Cuttack

Kendrapada

Kandhamal

Nayagarh

Jagatsinghpur

Khurda

Puri

Bhetnoi Area
(Ganjam District)

Ganjam

Balukhand Konark Area
(Puri District)

Bay of Bengal

RESEARCH

A research scheme was partially implemented with financial assistance from the Ministry of Environment and Forests, Government of India for collection of basic scientific information on the isolated population of Blackbuck in Balipadar-Bhetnoi area during 1995-96. On the findings of the above study there has been documentation of (*i*) population status, sex composition, herd structure and social grouping (*ii*) distribution pattern of the animal and threat aspects, and (*iii*) breeding biology (natality/mortality parameters), etc.

REHABILITATION

Steps were taken to rehabilitate Blackbuck in Bhitarkanika Wildlife Sanctuary during 1985-87 by introducing 14 (9M+5F) zoo bred specimens of Nanadanakahan Zoological Park stock. They could not adjust to the new surrounding and all perished after a couple of months.

MORPHOLOGY

Blackbuck is a medium sized Antelope which stands about 80 cm. at the shoulder and weighs about 40 kg. They are sexually dimorphic. The males at their initial stage are brown without horns. However, with secretion of sexual hormone, males develop a pair of unbranched, 'corkscrew' horns one on each side of head and change their body colour to elegant black. The beautiful spiral horns that never shed like deer's antler may grow upto 50 cm. The colour of the body coat is light yellow in young and females. As the male grows older the dorsal body colour turns into black. Males have pronounced post orbital glands which exude a pungent sticky secretion.

HABITAT

It primarily covers three Forest Ranges *i.e.*, Buguda and Aska in Ghumsur south Division and Khallikote Range of Berhampur Forest Division. The state highway from Khurda to Berhampur via Nayagarh also passes through their habitat.

The Blackbuck habitat covers about 60% cultivated lands/cropped fields, 15% rocky elevations, 10% man made houses and roads, 8% forest covers, 5% water bodies and 2% horticulture farms and waste lands.

FOOD

Blackbucks live on fresh tender leaves, grass, crops, cereals, vegetables and leaves of shrubs and trees. They feed for a long time, select succulent grasses, tender shoots of crops and plants which help them to maintain water balance in their bodies. They can survive without drinking water from a day to a week.

CENSUS OF BLACKBUCKS

The Forest Department conducted a census of Blackbucks on 14.5.1973 in the Balipadar-Bhentnoi area. Subsequently, the census was conducted in the year, 1980, 1998 and 2007 in the same area. In order to ascertain the population, the survey area is divided into small segments and enumerators in each segment make total count of the animals from direct sighting.

Results of 2007-census indicate that, there is 40% increase in the Blackbuck population over the last count (2004 census), and out of three Forest Ranges, namely Buguda, Aska and Khallikote, Buguda Range alone holds 58% of the Blackbuck population at present.

Table 1: Population trend in Ballipadar-Bhetnoi area (1973-2007) showing total number (and % out of total).

Year	Male	Female	Total	Sext ratio Male: Female
1973	152 (29.1)	69 (13.2)	523	1:2.0
1980	129 (26.6%)	72 (14.8%)	485	1:2.2
1998	94 (17.1%)	81 (14.7%)	551	1:4.0
2004	212 (27%)	87 (11%)	786	1:2.3
2007	306 (27.8%)	131 (11.9%)	1101	1:3.6

MOVEMENT PATTERN

Blackbucks are gregarious in nature. Their movement depends upon factors like availability of fresh **vegetation,** availability of **water, human interference,** interference by **domestic animals** within their activity limit, and **environmental parameters** like temperature, wind and rainfall play important role for determining the movement pattern of Blackbuck.

HERD STRUCTURE AND SOCIAL GROUPING

The social organizations of Blackbucks are categorized into the following groups:

Mixed herd or **territorial herd** with one territorial male and females of all age groups.

Bechelor herd of all male member.

Herd of all female groups.

Lone adult male (wandering).

BREEDING

Blackbucks breed in all seasons but main rut takes place between February to May. The gestation period is about 5-6 months. Usually only one young is born at a time. Females of about two years old and above give birth to young ones. At the Nandanakanan Zoo in Orissa, a female fawned for the first time at the age of 2 years and one month and another at the age of almost two and a half years.

MORTALITY

The normal life span of Blackbuck is about 12 to 15 years. The maximum age recroded was 16 years and 10 months.

PREDATION

The young ones fall prey to a number of predators such as wolves, hyenas, jungle cats, jackals, pythons, wild pigs and feral dogs, etc.

Fig. 3.   Herd of Blackbucks with a lone male

Recommedations:

1. Prior to the amendment of Wildlife (Protection) Act. 1972 in 1991, Balipada-Bhetnoi area, comprising 64.21 sq. km in Ganjam district was declared as Game Reserve on 19.9.1989 Under Sec. 36 of the Act for giving protection to the Blackbucks and their habitat. Since the relevant Section has been deleted in the amended Act, there is need for consideration to give it the status of "Community Reserve" so that Blackbuck can be better protected under the provision of the Wildlife (Protection) Act, since the Species is in Schedule-1 of the Act.

2. A long term research programme on Man-Blackbuck association and conflict, besides studying biological and ecological aspects need to be taken up with the funding support from the Ministry of Environment and Forests, Govt. of India.

3. Night patrolling should be strengthened by the Forest Department to prevent poaching of Blackbuck by outsiders, who primarily target males(bucks) taking advantage of villagers deep sleep after a day's hard work.

4. The practice of Departmental cultivation of ragi, gram, etc. at strategic locations as fodder for Blackbucks should be encouraged. This practice will provide some relief to farmers.

5. In view of the Blackbuck damage to crops, it is advisable to get the farmer's crops insured with Forest Department contributing a part of the premium as a goodwill gesture.

6. Eco-development activities should be taken up for getting cooperation of the villagers which will help in protecting the Blackbuck. This programme should be taken up in and around the Blackbuck habitats.

7. A thorough awareness programme should be created among the people of all age groups including farmers and students through distribution of booklets, stickers, posters, T-shirts, caps and also through media (All India Radio and Doordarsan, etc.).

8. Capture of fawns (a few hours to days old) by some local people for pets proves to be most disastrous for future survival of the Blackbuck population. More than 80% of the captive population die before reaching yearling stage. This is completely illegal and against the conservation objectives. All possible measures should be taken to stop this cruel activity. The local people/ villagers should be motivated not be capture the newly born fawns. Any violation should be severely dealt with.

CONCLUSION

This animal is an object of special adoration for the Bishnoi community of western Rajastan and the Vala Rajputs of Saurashtra, besides the people of Balipadar-Bhetnoi area in Ganjam district and the Odisha. These communities/people have been enthusiastically protecting Blackbucks which are associated with their past history, folklore and religious sentiments. These people still continue to protect Blackbucks in the vicinity of their habitations, despite the damage that the Blackbucks cause to their agricultural crops.

Realizing the significance of Blackbucks, the Forest and Environment Department, Govt. of Odisha should take early step to declare the Balipadar-Bhetnoi habitat as a "Community Reserve" in accordance with the recent amendment of Wildlife (Proctection) Act.

ACKNOWLEDGEMENTS

I am thankful to the Chief Wildlife Warden, Odisha for the support to conduct study on this species in the Balipadar and Bhetnoi areas of Ganjam district. Mr. Deepak Ranjan Behera, GIS specialist and Mr. Santosh Kumar Sundaray, Computer Assistant assisted me during my study. I am thankful to them.

REFERENCES

Acharjyo, L.N. and Mishra, R. (173). A note on age of sexual maturity of two species of antelopes in captivity. J. Bombay nat. Hist. Soc. 70 (2): 378.

Anon (1993). Species. Newletter of the Species Survival Commission, IUCN, Gland, Switzerland. 20:41.

Daniel, J. C.(1967). Point Calimere Sanctuary. J. Bombaynat. Hist. Soc. 64(3):512- 523.

Dora, M. C. (1916). Prajamitra, Aska, ganjam, Orissa. 1(7):4.

Kar, S. K. (2000a). Study on " Population, habitat preference, feeding and survival of Blackbuck(*Antilope cervicapra*) in Balipadar-Bhetnoi Wildlife Reserve of Ganjam district, Orissa". Final technical report submitted to the Ministry of Environment & Forests (RE Section), Govt. of India, pp.1-52.

Kar, S.K.(2000b). Survival fo Blackbuck, *Antilope cervicapra* in Ganjam district of Orissa: an epitome of human-animal. The Twilight-Journal of Pugmarks, Kolkata. 2(2&3): 28-30.

Kar, S. K. and Panda, R.M. (1996). Indian Antelope. NWCSO Newsletter (May-June, 1996). p. 4.

Kar, S.K. and Panda, R.M. (1997). Blackbuck an Indian antelope. Orissa Review. 53 (10-11): 38-39.

Kar, S.K. (2001). Balipadar's Blackbuck: an insight in to myth and reality of human-blackbuck relationship.Published by State Wildlife Wing, Forest Department, Govt. of Orissa, Bhubaneshwar. pp.1-41.

Mohanty, S.C.; Kar, C.S.; Kar,S.K.and Singh, L.A.K. (2004); Wild Orissa. Wildlife Organisation, Forest Department, Govt. of Orissa. Bhubaneswar, pp. 1-82.

Mishra, Ch.G.; Patnaik, S.K.; Sinha, S.K.; Kar, C.S. and Singh, L.A.K. (1996). Wildlife Wealth of Orissa (Orissa Forest Department Publ.). Orissa Govt. Press, Cuttack, pp.1-185.

Patnaik, S.K. and Acharjyo, L.N.(1985). Wildlife Conservation in Orissa. Cheetal, 27:38-44.

Prasad, N.L.N.S. (1978). Territory in Indian Blackbuck, Antilopecervicapra. J. Bombay nat. Hist. Soc. 86(2):187-191.

Rajitsinh, M.K. (1989). The Indian Blackbuck. Natraj Publishers, Dehradun, pp. 1-155.

Schaller, G.B. (1967). The Deer and the Tiger, Chicago University Press, Chicago, pp. 149-173.

ZSI (1994). The Red Data book of Indian Animals (Part-1, Vertebrate), Zoological Survey of India, Culcutta (Kolkata), pp. 195-197.

❑❑❑

18

THE BREEDING BIOLOGY OF RED WATTLED LAPWING *VANELLUS INDICUS INDICUS* (BODDAERT)

Vasudha Chaudhari

ABSTRACT

Birds share with mammals, the distinction of being the most recent vertebrates to inhabit the eath. They appear to be a climax group at the height of their evolutionary process and are known to perform a variety of functions in the natural ecosystem. The survival and propagation of these feathered bipeds is very essential for the sustenance of life on earth. The present study aims to record the breeding biology of Red Wattled Lapwing *Vanellus indicus indicus*. For the purpose of study, the parameters taken are breeding season, nest construction, clutch size, egg laying pattern, incubation period, hatching pattern, nestling period, share of sexes in nest building, incubation and feeding the young ones. This study will shed light on the reproductive behaviour of this bird in the study area and help gather the information that could be used for planning future conservation and propagation of the species.

Keywords: *Breeding biology, Vanellus indicus indicus, clutch size, incubation period, hatching pattern, nestling period.*

INTRODUCTION

Birds are among the most familiar animals of our environment, and through the years, their activities have been observed with keen interest. They have been viewed as the indicators of environmental quality. They share with mammals, the distinction of being the most recent vertebrates to inhabit the earth. They appear to be a climax group at the height of their evolutionary process. The present study attempts to documents the breeding behaviour of *Vanellus indicus indicus*, commonly known as Redwattled Lapwing, a resident bird in the study area. It is a large, brown, white and black bird with bright yellow long legs, red wattle and red bill with a black tip. It is found singly, in pairs or in small flocks of 10 to 20 individuals either in the fields or on the banks of rivers or streams or on a rock lying in a water body, standing motionlessly or preening their body parts or feeding. They have also been observed sitting inside the water bodies enjoying bathing. When approached, the birds fly off immediately, giving out series of loud shrill calls. Their flight is short and the birds soon settle down again. The birds feed on the ground only and have never been observed perching on the trees.

MATERIAL AND METHODS

The study has been carried out in Jammu district which has a total area of 3097 sq. Km. It lies between latitude 32°-50" and 33°-30" North and longitude 74°-24" and 75°-18" East. Height of Jammu District from mean sea level is 340 m-410 m. The climatic conditions in and around the study area are dry, sub-humid to arid. The average rainfall ranges from 100-120 cm.

For studying the breeding biology, attention has been focused on breeding season, nest construction, clutch size, egg laying pattern, incubation period, hatching pattern, nestling period, share of sexes in nest building, incubation and feeding the young ones. Incubation period was calculated from the date of laying of the last egg of the clutch to the hatching of that egg. In the same way, nestling period has been recorded from the date of hatching out of the last chick up to its fledging time. The nests were watched at fixed hours during the laying and hatching periods at 7.00-8.00 hours, 12.00-13.00 hours and 17.00-18.00 hours. Several irregular visits were also made during different hours of the day to observe the activities of the parent birds. Photographs were also taken during these visits.

Numerical Interpretations

1. Numerical values outside parenthesis against a particular feature include the number of specimens in which the feature is recorded. For instance, in case of clutch size 4 (5), 3 (4), 2 (3) would imply that five nests out of twelve have 4 eggs each, 4 nests are with three eggs each and three nests of two eggs each.
2. N is the total number of specimens in which that feature has been seen.
3. All egg lengths are in millimetres.

OBSERVATIONS

The observations recorded in the study area regarding the breeding behaviour of *Vanellus indicus indicus* are as under:

Breeding Season: April to July

Nest: Nest is a natural depression in the ground or on the roof tops encircled with pebbles of hard clay or even without a cover. It is always placed in an area from where the source of water is reasonably nearby.

Out of the four nests studied, two were made on the ground, one near a stream and one nest was made on the roof top having water tanks.

Eggs: Clutch size: N= 4, range 4 eggs; Shape of the eggs- Pyriform; Colour of the eggs – Pale olive green blotched with black; Size of the eggs- 42.2- 45mm x 31.6- 34.9 mm; Weight of the eggs- 17.5 gm- 19 gm.

Egg laying pattern: N= 4, eggs laid at an interval of 48 hours.

Incubation period: N=3; range 29 to 30 days; 29 (1) 30 (2). One nest out of the initially observed four nests had been destroyed at the egg level.

Hatching pattern: N= 3, three chicks hatched on the same day although at different hours and the remaining one hatched out after an interval of 48 hours. This shows that the incubation started after the laying of three eggs.

Nestling period: Chicks left the nest on the day of hatching but remained with their parents for 5-6 weeks. The newly hatched chicks are nidifugous and are capable of feeding themselves. The upper parts of the newly hatched chicks are grayish brown mottled with black and the lower parts are white. Beak is black in colour and the legs are dirty cream. Both the eggs and the chicks are superbly camouflaged. When the chicks are approached, the parents start giving out alarm calls. On hearing these calls, the chicks either hide themselves behind a stone or in a hole or lie listlessly on the ground becoming completely undetectable. On the disappearance of the intruder, the parents again give out a series of high pitched calls, and the chicks on receiving such type of calls, get up and resume their feeding activity.

Share of Sexes: Both the parents take part in nest building and incubation. It has been observed that when the nest is approached, the incubating bird does not fly but walks away from the nest keeping its head and neck at the level of the chest so as to escape attention.

DISCUSSION

In the study area, the eggs were laid at an interval of 48 hours. The present study confirms the observations of Desai and Malhotra (1976) as far as the laying pattern in the 3 nests as reported by them is concerned. In the fourth nest, under their study, having clutch size of 3, two eggs were laid at an interval of 48 hours and the third one was laid at an interval of 96 hours after the laying of the second egg. Such a type of laying pattern has not been recorded in the study area.

Incubation in all the nests observed in the study area started after the laying of 3 eggs, but Desai and Malhotra *(op. cit.)* have reported it to start after the laying of the first egg.

The behaviour of the incubating bird when the nest is approached in the study area is in agreement with the findings of Naik, *et.al.* (1961).

Redwattled Lapwing, *Vanellus indicus*
Incubating the egg

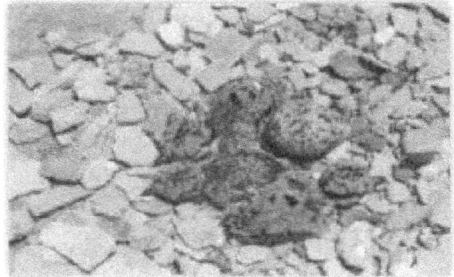

1 egg and 3 newly hatched chicks of
Redwattled Lapwing, *Vanellus indicus*

REFERENCES

Ali, S. and Ripley, S. D. (1968-74). *The Handbook of Birds of India and Pakistan*; 10 vols. Oxford University Press, Bombay.

Bates and Lowther, E.H.N. (1952*Breeding Birds of Kashmir*, Oxford University Press, Bombay.

Desai, J. H. and Malhotra, A.K. (1976). " A note on incubation period and reproductive success of the Redwattled Lapwing, *Vanellus indicus* at Delhi Zoological Park"; J. Bombay Nat. Hist. Soc., 73, pp. 392-394.

George, N.J. (1958). "On the parental care of Yellowwattled Lapwing, *Vanellus malabaricus*" ; J. Bombay Nat. Hist. Soc., 82, pp. 655-656.

Jamdar. N.(1985). " Redwattled Lapwing, (*Vanellus indicus*), suffering from cataract"; J. Bombay Nat. Hist. Soc., 82, pp. 197.

Naik, R.M., George, P.V. and Dixit, D.B. (1961). "Some observations on the behavior of the incubating Redwattled lapwing *Vanellus indicus indicus* (Bodd.)", J. Bombay Nat. Hist. Soc., 58, pp. 223-230.

Vyas, R. (1997). "Flocking and courtship display in Redwattled lapwing (*Vanellus indicus*)"; J. Bombay Nat. Hist. Soc., 94, pp. 406-407.

❑❑❑

19

WETLAND AVIFAUNA OF KUNDAVADA LAKE, DAVANAGERE DISTT., KARNATAKA

M.N. Harisha, B. Hosetti and Shahnawaz Ahmad

ABSTRACT

This paper documents a list of wetland birds with reference to the migratory birds of Kundavada Lake, Davanagere Dist, Karanataka. Till date, there is no report on the wetland birds of Kundavada Lake and the present study is therefore first of its kind. Hence, the study becomes the preliminary data for future investigation and we sighted 53 species of birds belonging to 16 families.

Keywords: *Wetlands, Birds Diversity, Kundavada Lake, Karnataka.*

INTRODUCTION

Wetlands have been described as a halfway world between terrestrial and aquatic ecosystems that exhibit some of the characteristics of each (Wanger, 2004). They form part of a continuous gradient between uplands and open water. Wetlands have many distinguishing features, the most notable of which are the presence of water, unique soils and vegetation, adapted to or tolerant or saturated soils (Miller, 2000). According to Coward *et al.* 1979, " Wetlands are areas of marsh, fen, peat land or water, whether natural or artificial, flowing, fresh brackish or salt, including areas of marine water, the depth of which at low tide does not exceed six meters."

Wetlands are specialized ecosystems which perform important ecological function and have many ecological, socio-economic and cultural values. Wetlands are known to be the most productive and diverse ecosystems on earth because they provide direct benefits to people as sources of food, recharge of aquifers, regulation of water quality, natural purification of waste water, reducing sediment load, water recharge, recycling of bio-genic salts, as a source of agricultural water, animal husbandry, agriculture and also as a refuge for rare and endangered species of plants and animals (Hosetti, 2002). Wetlands preserve genetic and community diversity and provide food and habitat for migratory birds and other creatures. Birds are among the most eye-catching of wetland animals and various species are extremely sensitive to large hydrological changes (Crowder and Bristow, 1988). The present study aims to document the checklist of aquatic avifaunal diversity of Kundavada Lake, Davanagere since no avifauna studies have been carried out previously on the birds prevailing in this area.

STUDY AREA

The Kundavada Lake is a spectacular and marvelous site for avifauna diversity; it is located between latitude of N 14° 27' 30" and longitude of E 75° 53' 39". This wetland provides water for drinking to Davanagere city. The lake is about 243.27 acres and just near to the Pune-Bangalore highway. The lake is an attracting sight for many wetland birds as it supports good nesting site and is full of aquatic flora including *Ipomea* species (sp.), *Bergia* sp., *Salvinia* sp., *Ceratophyllum* sp., *Hydrilla* sp., *Alternanthara* sp. and many more. The vegetation across the lake consists of Neem, Coconut, *Lantana* sp., *Calotrophis* sp., *Caesalpinia* sp., *Bauhinea* sp. and grasses. Animal food sources include Phytoplankton, Zooplankton, Pisces, Molluscans and Insects. The lake has a water connection from the Channel of River Tungabhadra. Lake is free from sewage and agricultural drainage. It is also a good recreational place for the people and the students to conduct research. The place is used for commercial game fishing also to avoid the disturbances for birds.

MATERIAL AND METHODS

The checklist prepared is based on the field work conducted during October 2007-September 2008 across Kundavada Lake by foot method i.e. road side count (Burnham *et al.*, 1980; Simpson 1949). A total of 12 visits (1 visit per month) were made in the field for observing the bird diversity. Birds were obsereved from 6 a.m. to 11 a.m. and identified using Olympus binoculars (10×50) and field guides of Ali *et al.* (1983); Grimmett *et al.* (2001). In the heronry, total counts were carried out by direct and point counting methods for the birds. The nomenclature followed here was given by Manakadan and Pittie (2001). The status on the movement and seasonality of occurrence, the parameters are listed as; LM-Local migratory, WM-winter migratory and R-Resident depending on movement and seasonality (Table 1).

Table 1. Checklist of wetland birds of Kundavada Lake, Davanagere district, Karnataka with common/vernacular, scientific names, their status and abundance.

S.No.	Common Name	Scientific Name	Status	Frequency
Order: Pelecaniformes **Family: Phalacrocoracidae**				
1.	Little Grebe	*Tachybaptus ruficollis* (Pallas)	LM	Common
Order: Pelecaniformes **Family: Phalacrocoracidae**				
2.	Little Cormorant	*Phalacrocorax niger* (Vieillot)	WM	Common
3.	Indian Shag	*Phalacrocorax fuscicollis* (Stephens)	WM	Rare
Family: Anhingidae				
4.	Darter	*Anhinga melanogaster* (Pennant)	R	Common

Contd...

Table 1. Contd...

S.No.	Common Name	Scientific Name	Status	Frequency
Order: Ciconiiformes **Family: Ardeidae**				
5.	Little Egret	*Egretta garzetta* (Linnaeus)	R	Comman
6.	Gery Heron	*Ardea cinerea* (Linnaeus)	R	Rare
7	Purple Heron	*Ardea purpurea* (Linnaeus)	R	Rare
8	Black-crowned Night Heron	*Nycticorax nycticorax* (Linnaeus)	R	Rare
9	Median Egret	*Mesophoyx intermedia* (Wagler)	R	Common
10	Cattle Egret	*Bubulcus ibis* (Linnaeus)	R	Common
11	Indian Pond Heron	*Ardeola grayii* (Sykes)	R	Common
Family: Ciconiidae				
12	White-necked Stork	*Ciconia episcopus* (Boddaert)	R	Rare
Family: Threskiornithidae				
13	Oriental White Ibis	*Threskiornis melanocephalus* (Latham)	R	Common
14	Black Ibis	*Pseudibis papilosa* (Temminck)	R	Common
15	Glossy Ibis	*Plegadis falcinellus* (Linnaeus)	WM	Rare
Order: Anseriformes **Family: Anatidae**				
16	Lesser Whistling Duck	*Dendrocygna javanica* (Horsfield)	LM	Abundant
17	Bar-headed Goose**	*Anser indicus* (Latham)	WM	Rare
18	Brahminy Shelduck**	*Tadorna ferruginea* (Pallas)	WM	Occasional
19	Spot Billed Duck	*Anas poecilorhyncha* (J.R. Forester)	R	Common
20	Cotton Teal	*Nettapus coromandelianus* (Gmelin)	R	Common
21	Marbled Teal**	*Marmaronetta angustirostris* (Menetries)	WM	Rare
22	Gadwall**	*Anasstrepera* (Linnaeus)	WM	Occasional
23	Eurasian Wigeon**	*Anas Penelope* (Linnaeus)	WM	Occasional
24	Northern Shoveler**	*Anas Clypeata* (Linnaeus)	WM	Common
25	Northern pintail**	*Anas acuta* (Linnaeus)	WM	Common
26	Garganey**	*Anas querquedula* (Linnaeus)	WM	Abundant
27	Common Pochared**	*Aythya ferina* (Linnaeus)	WM	Abundant

Contd...

Table 1. Contd...

S.No.	Common Name	Scientific Name	Status	Frequency
Order: Gruiformes **Family: Rallidae**				
28	White- breasted Waterhen	*Amaurornis phoenicurus* (Pennant)	R	Common
29	Common Moorhen	*Gallinule chloropus* (Linnaeus)	R	Common
30	Purple Moorhen	*Porphyrio porphyrio* (Linnaeus)	R	Abundant
31	Common Coot**	*Fulica atra* (Linnaeus)	WM	Abundant
Order: Charadriiformes **Family: Jacanidae**				
32	Pheasant-tailed Jacana	*Hydrophasians chirurgus* (Scopoli)	R	Abundant
33	Bronze-winged Jacana	*Metopidius indius* (Latham)	R	Abundant
Family: Rostratulidae				
34	Greater Painted Snipe	*Rostratula benghalensis* (Linnaeus)	R	Occasional
Family: Charadriidae				
35	Little Ringed Plover**	*Charadrius dubius* (Scopoli)	WM	Common
36	Yellow- Wattled Lapwing	*Vanellus malabaricus* (Boddaert)	R	Common
37	Red- Wattled Lapwing	*Vanellus indicus* (Boddaert)	R	Common
Family: Scolopacidae				
38	Black- tailed Godwit	*Limosa limosa* (Linnaeus)	R	Common
39	Spotted Redshank**	*Tringa erythropus* (Pallas)	WM	Rare
40	Common Redshank**	*Tringa tetanus* (Linnaeus)	WM	Rare
41	Marsh Sandpiper**	*Tringa stagnatilis* (Linnaeus)	WM	Common
42	Common Greenshank**	*Tringa nebularia* (Gunner)	WM	Rare
43	Green Sandpiper**	*Tringa ochropus* (Linnaeus)	WM	Rare
44	Common Sandpiper**	*Tringa hypoleucos* (Linnaeus)	WM	Common
45	Little Stint**	*Calidris minuta* (Leisler)	WM	Common
Family: Recurvirostridae				
46	Black- winged Stilt	*Himantopus himantopus* (Linnaeus)	R	Common

Contd...

Table 1. Contd...

S.No.	Common Name	Scientific Name	Status	Frequency
Family: Larodae				
47	River Tern	*Sterna aurantia* (J.E. Gray)	LM	Common
Order: Coraciiformes **Family: Alcedinidae**				
48	Small Blue Kingfisher	*Alcedo atthis* (Linnaeus)	R	Common
49	Lesser Pied Kinfisher	*Ceryle rudis* (Linnaeus)	R	Common
50	White-breasted Kingfisher	*Halcyon smyrnensis* (Linnaeus)	R	Common
Order: Passeriformes **Family: Motacillidae**				
51	Yellow Wagtail**	*Motacilla flava* (Linnaeus)	WM	Rare
52	Grey Wagtail	*Motacilla cinerea* (Tunstall)	WM	Common
53	Large Pied Wagtail	*Motacilla maderaspatentsis* (Gmelin)	R	Common

Note:

Status of the birds observed : Resident (R), Winter Migratory (WM), Local (resident) Migaratory (LM)

**Birds with migratory population.

RESULTS AND DISCUSSION

The study reveals the occurrence of 53 species of birds belonging to 16 families and 8 orders form the study area (Table 1). The details such as common and scientific names, status and abundance of the wetland birds are presented in Table 1. The order Charadriiformes dominated the list by 6 families with 16 species followed by order Anseriformes with 12 species, 3 Ciconiiformes families with 11 species, Gruformes with 4 species, Coraciiformes, Passeriformes, Pelecaniformes with 3 species each and Podicipediformes with 1 species respectively. Out of total 53 species, 27 were resident, 20 were winter, 3 were local and rest 3 were resident migratory birds recorded. Most of the migratory species were winter visitors except Cotton Teal (*Nettapus coromandelianus*) and Lesser Whistling Duck (*Dendrocygna javanica*) which were summer visitors. Based on the frequency of sightings of avifauna of the Kundavada lake, Little Grebe (*Tachybaptus ruficollis*), Little Cormorant (*Phalacrocorax niger)*, Darter (*Anhinga melanogaster)*, Little Egret (*Egretta garzetta*), Median Egret (*Mesophoyx intermedia*), Cattle Egret (*Bubulcus ibis*) and Indian Pond-Heron (*Ardeola grayii*), Oriental White Ibis (*Threskiornis melonocephalus*), Black Ibis (*Pseudibis papillosa*), Spot Billed Duck (*Anas poecilorhyncha*), Cotton Teal (*Nettapus coromandelianus*) Northern Shoveller (*Anas clypeata*), Northern Pintail (*Anas acuta*), White-Breasted

Water Hen (*Amaurornis phoenicurus*), Common Moorhen (*Gallinula chloropus*), Little Ringed Plover (*Charadrius dubius*), Red-Wattled Lapwing (*Vanellus indicus*), Yellow- Wattled Lapwing (*Vanellus malabaricus*), Black- tailed Godwit (*Limosa limosa*), Marsh Sandpiper (*Tringa stagnatilis*), Common Sandpiper (*Tringa hypoleucos*), Little Stint (*Calidris minuta*), Black- Winged Stilt (*Himantopus himantopus*), River Tern (*Sterna aurantia*), Small Blue Kingfisher (*Alcedo atthis*), Lesser Pied Kingfisher (*Ceryle rudis*), White-Breasted Kingfisher (*Halcyon smyrnensis*), Grey Wagtail (*Motacilla cinerea*), Large Pied Wagtail (*Motacilla maderaspatensis*), were the common and dominated species inhabiting 55% of these ponds/water bodies (Fig. 1), while Indian Shag (*Phalacrocorax fuscicollis*), Gery Heron (*Ardea cinerea*), Purple Heron (*Ardea purpurea*), Black-crowned Night Heron (*Nycticorax nycticorax*), White-necked Stork (*Ciconia episcopus*), Glossy Ibis (*Plegadis falcinellus*), Bar-headed Goose (*Anser indicus*), Marbled Teal (*Marmaronetta angustirostris*), Spotted Redshank (*Tringa erythropus*), Common Redshank (*Tringa tetanus*), Common Greenshank (*Tringa nebularia*), Green Sandpiper (*Tringa ochropus*), Yellow Wagtail (*Motacilla flava*) were rarely sighted with 24%. While Lesser Whistling Duck (*Dendrocygna javanica*), Garganey (*Anas querquedula*), Common Pochard (*Aythya ferina*), Purple Moorhen (*Porphyrio poryhyrio*), Common Coot (*Fulica atra*), Pheasant-tailed Jacana (*Hydrophasianus chirurgus*), Broze-winged Jacana (*Metopidius indicus*) were abundant with 13%. However Brahminy Shelduck (*Tadorna ferruginea*), Gadwall (*Anas strepera*), Eurasian Wigeon (*Anas Penelope*), Greater Painted Snipe (*Rostratuala benghalensis*) were occasionally sighted with 8% of frequency of occurrence (Fig. 1).

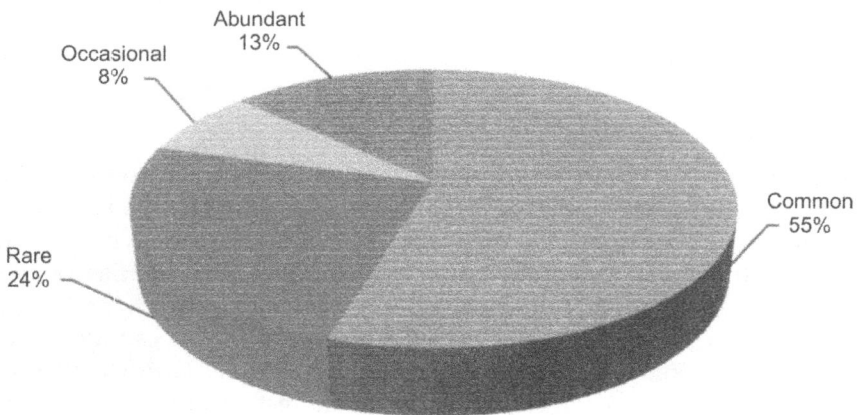

Fig. 1.   Frequnecy of occurrence of birds of Kundavada lake, Davanagere

Three globally threatened species were recorded, such as oriental white ibis (*Threskiornis melanocephalus*), oriental dartec (*Anthinga melanogaster*) and black-tailed godwit (*Limosa limosa*), These are listed in the near threatened category by IUCN in 2010. These water birds were found to utilize different wetland habitats extensively for foraging, nesting and roosting on the emergent and fringed vegetation. Water birds, being generally at or near the top of most wetland food chains are highly susceptible to habitat disturbances and are therefore good indicators of general condition of aquatic habitats (Kushlan, 1992; Jayson and Mathew, 2002; Kler, 2002). The rich diversity of the wetland birds documented during the present study may be because of availability of varied sources of food as well as foraging. The wetland birds are in general being heterogeneous in their feeding habits (Ali and Ripley, 1987). Thus wetland birds exploit a variety of habitats and depend upon a mosaic of microhabitats for their survival. Paddy fields with stray trees and scattered vegetation cover might have extened comfortable shelter and suitable foraging grounds for the wetland birds. This habitat by supporting different food sources like fish, crustanceans, invertebrates, water plants and planktons further add to the diversity of wetland birds (Basavarajappa, 2004). This indicates that the habitat is more suitable and supports all the visitor birds as well as resident birds by providing immense food enough space to breed. Every year from October onwards a considerable number of water birds reach the wetland. Highest bird density was recorded during winter months, when the anthropogenic activities are minimum. The peak winter population of the migratory birds was seen during the month of January and February. The basic requirement of the migratory water birds at their wintering sites are adequate food supply and safety (Bharat Lakshmi, 2006), almost all of them leave the wetland by March-end or early April.

CONCLUSION

The study documents the rich avifauna diversity showing the area still provides some potential habitats for the declining population of the threatened birds. Therefore, it is the need of the hour to monitor these areas systematically in the rapidly changing environment with a focused study on status, distribution and conservation of the avifauna of the region. This can be achieved only through strengthening public participation in the study of status, distribution and conservation of birds of Kundavada Lake, Davanagere district, Karnataka. The study could effectively provide the baseline for research and data which could be used for conservation purpose of avifauna diversity.

ACKNOWLEDGMENTS

We are thankful to the Department of PWD Government of Karnataka, India for permitting to undertake the study in Kundavada Lake, Davanagere. We also thank villagers residing around the lake for providing all the facilities during the study period.

REFERENCES

Ali, S. and Ripley, S. D. (1983). Hand Book of Birds of India and Pakistan. Oxford university Press, Delhi.

Ali, S. and S. D. Ripley (1987). Compact handbook of the birds of India and Pakistan together with those of Bangladesh, Nepal, Bhutan and Sri Lanka. Oxford University Press, Delhi.

Basavarajappa, S. (2004). Avifauna of agro-ecosysmtems of Maidan area of Karnataka. Zoos' Print J. 21 (4):2217-2219.

Bharata Lakshmi, B. (2006). Avifauna of Gosthani estuary near Visakhapatnam, Andhra Pradesh. J. Natcon., 18(2):291-304.

Burnham, K.P., Anderson, D.R. and J.L. Laake (1980). Estimation of density from line transect: Sampling of Biological populations. Dilak Monogs.72:202.

Crowder, A.A., and J. M. Bristow (1988). The future of waterfowl habitats in the Canadian lower Great Lakes wetlands. Journal of Great Lakes Research 14: 115-127.

Grimmet, R., C.Inskipp & T. Inskipp (2001). Birds of Indian Subcontinent Oxford University press. Delhi, p. 384.

Hosetti, B.B. (2002). Wetlands Conservation and Management, Pointer Publishers, Jaipur, India.

Jayson, E.A. and D.N. Mathew (2002). Structure and composition of two birds communities in the southern Western Ghats. J. Bombay Nat. Hist. Soc. 99(1):8-25.

Kler, T. K. (2002). Birds species in Kanjali Wetland. Tiger Paper 39(1):29-32.

Kushlan, J.A. (1992). Population biology and conservation of colonical water birds. Colonial Water Birds 15:1-7.

Manakadan, R. and A. Pittie. (2001). Standardized common and scientific names of the birds to the Indian Sub continent. Buceros, 6(1): 1-37.

Miller, G. T. (2000). Living in the environment. Brooks/Cole publishing company, California.

Mitch, W. J., and J. G. Gosselink (2000). Wetlands. John Wiley and Sons, Inc., New York.

Simpson, E.H. (1949). Measurement of Diversity Nature, 163:688.

Wanger, M. (2004). Managing riparian habitats for wildlife. Report No PWD BR W7000-306, Texas Parks and Wildlifes Department, Austin.

❏❏❏

www.ingramcontent.com/pod-product-compliance
Lightning Source LLC
Chambersburg PA
CBHW021434180326
41458CB00001B/271